REPLACEABLE YOU

ALSO BY MARY ROACH

Fuzz: When Nature Breaks the Law

Grunt: The Curious Science of Humans at War

Gulp: Adventures on the Alimentary Canal

Packing for Mars: The Curious Science of Life in the Void

Bonk: The Curious Coupling of Science and Sex

Six Feet Over: Science Tackles the Afterlife (originally published as *Spook*)

Stiff: The Curious Lives of Human Cadavers

MARY ROACH

W. W. NORTON & COMPANY
Independent Publishers Since 1923

REPLACEABLE YOU

Adventures in Human Anatomy

This book is intended as a general information resource; it is not a substitute for the advice of your health care provider. Neither the publisher nor the author can guarantee the complete accuracy, efficacy, or appropriateness of any recommendation in every respect.

Copyright © 2025 by Mary Roach

All rights reserved
Printed in the United States of America
First Edition

For information about permission to reproduce selections from this book, write to Permissions, W. W. Norton & Company, Inc., 500 Fifth Avenue, New York, NY 10110

For information about special discounts for bulk purchases, please contact W. W. Norton Special Sales at specialsales@wwnorton.com or 800-233-4830

Manufacturing by Lakeside Book Company
Book design by Chris Welch
Production manager: Devon Zahn

ISBN 978-1-324-05062-9

W. W. Norton & Company, Inc., 500 Fifth Avenue, New York, NY 10110
www.wwnorton.com

W. W. Norton & Company Ltd., 15 Carlisle Street, London W1D 3BS

10 9 8 7 6 5 4 3 2 1

For Otis

CONTENTS

First Thoughts   1

1   To Build a Nose
The Dawn of Replacement Body Parts   9

2   Gimme Some Skin
Replacing the Human Exterior   21

3   Mixed Meats
Humans with Pig Organs, and Pigs with Human Organs   35

4   Heart in a Box
Creating Ultra-Long-Life Organs   53

5   The Vagina Dialogue
Repurposing Your Parts   69

6   Giving the Finger
Some Transplants Are Tougher Than Others   81

7   The Cut-Off Point
Longing for a Prosthetic Leg   91

8   Joint Ventures
Woodworking Without Wood   111

9	**Intubation for Dummies**	
	The Brief Terrors of Mechanical Breathing	129
10	**Heavy Breathing**	
	Inside the Iron Lung	139
11	**The Mongolian Eyeball**	
	With Cataract Surgery, Sometimes Simpler Is Better	155
12	**The Last Six Inches**	
	Battling the Stigma of Ostomy	173
13	**Out of Ink**	
	How to Print a Human	189
14	**Shaft**	
	Hair Transplants Through the Ages	201
15	**Splitting Hairs**	
	Grow Yourself from Scratch!	213
16	**The Ass Men**	
	Chasing Perfection with Math and Fat	227
17	**Some of the Parts**	
	A Day in the Life of a Tissue Donor	243
	Last Thoughts	253
	Epilogue	259

Acknowledgments	261
Sources	264

REPLACEABLE YOU

First Thoughts

The Victorian upper crust excelled at taking apart dinner. Plates were flanked by as many as sixteen utensils. There were special knives and forks for game and roasts and separate sets for fish and salad and fruit. But tableware only takes one so far. To reduce a mouthful of mutton or hare to what is known in oral-processing parlance as "the swallowable state," a certain amount of chewing must take place. For that one needs teeth. For the multitude of Victorians who lacked them, a seventeenth implement was sometimes set out. Picture a pair of garden pruners with flat, toothed blades. The masticator, as it was called, allowed the edentulous dinner guest to pre-chew the offerings before lifting them to mouth.

Dentures have been around since the 1700s. Who would rely on a masticator when they could opt for false teeth? In fact, people often had both. They'd masticate in the privacy of their room, then install their dentures and head into the dining room to sit down and not eat. Far from being a functional replacement for molars and incisors, false teeth at the time were largely ornamental, a sort of wig for the mouth. "With uncertain adaptation to the jaws, and primitive fastening arrangements, false teeth were in general almost useless for eating," wrote the late John Woodforde in *The Strange Story of False Teeth*, a book I chewed through to here expectorate a few morsels.

At its most extreme, the unsatisfactory fastening arrangement took the form of piercings in the gums, from which a set of "floating teeth" were suspended "on the ear-ring principle," as Woodforde noted. Moderately less cruel were spring-loaded dentures, wherein stiff wire coils united the lowers and uppers, pushing the latter against the top gums, and at the same time tending to launch them forward out of the mouth. "It shoots beyond the gums and . . . forces the lip out just under the nose," George Washington complained in one of his many letters to one of his many dentists. "His mouth was like no other I ever saw," recalled a visitor, "the lips firm and the under jaw seem[ing] to grasp the upper with force, as if the muscles were in full action when he sat still." Hence the grim set of the general's mouth in his later portraits. He was locked in battle with his teeth.

The twentieth-century denture stayed put through a combination of suction and oral glue—or "fixative," as the Poligrip people prefer. Even then, dentures afforded less than a quarter of the chewing efficiency of natural teeth. Respondents to a 2015 Poligrip survey confessed that their dentures limited not only their menu choices but their love life. Twenty-four percent of them were afraid to kiss someone passionately. Twenty percent were reluctant to smile.

Given all this, the truly astonishing item in that Poligrip poll was that the majority of the respondents—53 percent—were younger than forty-five when they first got dentures. Meaning that most had made the decision to have pulled their remaining healthy teeth.

It called to mind something I'd heard about—"matrimonial dentures," wherein a bride-to-be* was given false teeth and the

---

* That it was only brides who had it done, never grooms, has given rise to dubious denture conspiracy theories, one of course involving "gum jobs" and another best related by the denture wearer herself. "All the women I know who wear dentures have trouble talking," wrote the author of a "Dear Abby" letter, "because the dentist intentionally made them that way so women will keep their mouths shut."

necessary extractions as a wedding gift, to spare the couple the expense of having rotted teeth extracted one at a time over a lifetime. (Likely a misguided strategy, as denture wearers had to be refitted for a new set every five to ten years, as the body resorbed the toothless bone below.) My understanding of the practice was that it was confined to a reasonably distant past and to insular, dentally underserved cultures that relied on the services of denturists, people who are not licensed to fill or clean or otherwise extend the lifespan of a tooth.

Looking into it, I landed on a Reddit r/AskHistorians thread. Someone had inquired about the practice. More than a thousand comments followed, from people whose parents or grandparents—in the United States, the UK, Canada, Australia, Germany—had had all their teeth pulled in their early twenties or teens to make way for dentures. At its peak, the practice spanned the 1940s, '50s, and '60s. Of the first one hundred comments, which was as far as I read, only eight referenced wedding gifts. Eleven were presents for youth who had just come of age (almost including Paul McCartney, who told Terry Gross in a 2013 *Fresh Air* interview that his dad had suggested that at twenty-one he "have all my teeth taken out and false teeth put in"). Mostly it seemed to be "just what you did," as one Redditor recalled their grandparent saying. "Commonplace," noted another. "Fashionable." "All the rage back then."

I suspected unscrupulous dentists may have pressured or misled these people's relatives, but I saw no mention of that. Fluoridation didn't hit its stride until the 1960s, and the absence of fluoride likely contributed to the poor state of people's teeth. But it seemed to me that at least some of these folks had succumbed to the lure of what they took to be progress, believing false teeth would not only look better than their natural teeth but would function equally well—that they'd serve as literal replacements for what they'd been born with.

At its most basic, progress is simple chronology: one thing following on the next. With fabricated body parts, the progression has followed a predictable course. The earliest offerings existed independent of the part they replaced: masticators, spectacles, wigs. Over time, devices migrated inside the body—masticators gave way to dentures, glasses to contact lenses, toupees to hair transplants. Replacements became more integrated, more complex, more expensive. False teeth and prosthetic limbs can now be screwed into bones, artificial lenses surgically implanted. Of late, the buzz is for regeneration: bits and pieces grown from one's own cells. Stem cells and gene editing have landed us on the brink of a medical revolution. When it comes to regenerating entire complex body parts, however, it is a wide brink, with plenty of open terrain for hype.

When I was a kid, there was a TV show called *The Six Million Dollar Man*. Every week, in the opening credits, test pilot Steve Austin would crash his top-secret supersonic aircraft into the tarmac. "We can rebuild him," says an unidentified, optimistic narrator. "Better than he was before. Better. Stronger. Faster." Cut to heavily sped-up footage of the actor in a red track suit, running, and then stopping to scan the horizon with his bionic eye, which, annoyingly, beeps when he cues it up. At the time, this was understood to be science fiction. Today it can feel like Steve Austin has arrived, like he's right outside your door, holding out his neuroelectric hand with articulated fingers, saying, *Come with me to this brave new world! Grow a kidney! Print a leg!* Headlines regularly serve up grabby pronouncements of inconceivable feats. "Lab-Grown Brain Cells Play Video Game Pong." "Dr. Canavero Performing World's First Head Transplant in December." "Lab-Grown Penises Are on the Horizon." (*Run!*)

Fast-forward a few years, and nothing more is heard. Are the penises still lurking out there on the horizon? Have the brain cells mastered *Super Mario*? Whatever became of the head transplanter?* For every step forward, three go nowhere. Progress doesn't march, it lurches. In the best of circumstances, it takes a decade for all the testing and finessing, the up-scaling and down-pricing, all the things that need to happen to bring something new into clinical practice.

If things haven't progressed as quickly or as smoothly as *The Six Million Dollar Man* led us to think they would, it's not because of researchers' shortcomings. As we'll see, their achievements are dazzling almost beyond comprehension. Blame instead the overwhelming complexity of the human body. It's tough for a few hundred years of medicine and engineering to compete with the evolutionary accomplishments of millions of years of natural selection.

By way of warning—or reassurance—this is not a roundup of the latest advances in regenerative medicine. I have neither the background nor the quick production turnaround for a book like that. I dive into stem cells and bioprinting and gene-edited pig parts, but what I offer is more of a primer, a reality check for those who, like me, find themselves bobbing along in the swift current of discoveries that feel at once wondrous, improbable, and surreal.

Nor am I here to predict the future. There are no cyborgs clanking through these pages. Like all my books, this one situates itself for the most part in the present. It was an opportunity,

---

*   Still making widely discredited claims, though now about the feasibility of brain transplants rather than head transplants, lest, in his words, the "aged face and other head tissues . . . defeat the purpose of enjoying a pristine body." Also dabbling in self-help publishing. Here's the Amazon summary of Canavero's 2013 advice book for men: "This book will tell you . . . how to exploit the needs and weaknesses of women for your benefit, how to cajole them and lure them into your web . . ."

a two-year hall pass, to spend time with people and in worlds I would otherwise never have access to. I'm drawn to the human elements of the quest. How does a person—and their surgeon—decide that it's time to cut off an underperforming foot and replace it with a prosthetic? How do you combat the stigma of ostomy? How do you remove a tissue donor's bones in a way the family will be comfortable with? Things like that. Also, things like this: Could a heart survive indefinitely outside a body? Is a finger a workable substitute for a penis? What makes a pig a better organ donor than a goat?

In the course of reporting this book I spent some time with a surgeon whom you'll meet, Jeremy Goverman. At one point, I had mentioned the title of my book. Toward the end of our second day together, apropos of I forget what, Goverman turned to me. "I don't think you *can* replace the human body," he said. But it sure is interesting to try.

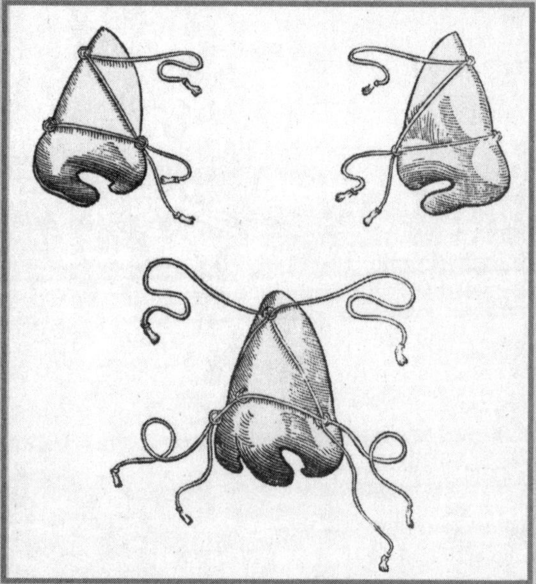

1

To Build a Nose

The Dawn of Replacement Body Parts

Tycho Brahe's nose came off in 1566. By all accounts, the celebrated astronomer was a brash and hot-headed individual. An argument at a gathering at a professor's home—Brahe still a student at the time—led to a late-night duel of honor. As custom then held, the duelists took up swords. A rapier swipe removed the better part of Brahe's nose, leaving the nasal cavity open to view. For the rest of his life, Brahe used a metal nosepiece, likely brass, painted to match his complexion and glued in place. Though not very well. "Occasionally," writes one biographer, "it would drop off." Along with whatever he used to shape his swooping red mustaches—their wingspan in some portraits stretching beyond the perimeter of his ruff collar—Brahe carried with him a small box of nose adhesive.

It would be another few centuries before prosthetic noses reliably stayed put. Some of the improvements came down to advances in materials science. Lighter options became available—aluminum, vulcanized rubber, celluloid plastic. New methods were devised to secure the nose. Here is onetime army surgeon Frank Tetamore describing one such invention—his own—in an 1894 paper: "These artificial noses are made of a very light plastic material. . . . They are secured on the face by bow spectacles made especially for the purpose." To obscure the lower border of the prosthesis, "a mustache [was] fastened to the nose piece." Forty years before novelty companies began selling Groucho Marx glasses, Frank Tetamore had invented a medical version. (Cigar to be avoided. The plastic he used, celluloid, is highly flammable.) The noses were finished with oil paint, different colors deemed suitable for different types of light. "He has one which he wears during the day," wrote Tetamore of a prominent New York merchant who became a patient, "and one for evenings."

Advancements in prosthetic nose technology paralleled those in the realm of artificial teeth. The requirements were similar: a readily molded material that would not disintegrate or smell bad and, most importantly, a way to keep the item from shifting or dropping out of place. As an alternative to adhesives, the newer prosthetic noses could be spring-loaded, along the lines of eighteenth- and nineteenth-century dentures.

And so it makes sense that the inventor of the spring-loaded prosthetic nose was a dentist, Robert Upham. "My idea in the beginning was to see if I could not relieve the necessity for glasses on the nose," Upham wrote in the *Boston Medical and Surgical Journal* in 1901. As with the dentures, the wearer compressed the springs while donning the prosthesis. Once the nose was in position, the springs were released, allowing them to sit snug against what remained of the walls of the nostrils. "They are never painful," Dr. Upham insisted. (Securing a set of teeth of course demanded more aggressive springs. The force needed to hold the upper plates in position tended to push apart the jaws. Keeping one's mouth shut required mild, sustained effort—something Tycho Brahe might have benefited from.)

Of the many materials one might use to build a replacement nose, none surpasses one's own skin. Danish physicians in Brahe's day were not known to undertake nose rebuilds, though in fact the procedure had been around for some three thousand years. Rhinoplasty was the original plastic surgery.

There has long been a call for noses. As far back as 1500 BC, in India, and extending through the Roman Empire and ninth-century Ireland, nasal mutilation was a form of punishment. Because of the organ's visibility, sitting as it does in the middle of the face, nasal disfigurement served as both humiliation and a warning to the populace. Noses were hacked off for thievery, tax evasion, adultery, disloyalty. Bounties were placed on enemy noses. Entire towns denosed.* Syphilis brought renewed demand. In its later stages, the disease can cause the bridge of the nose to deteriorate and collapse.

Necessity—and billing potential—was the mother of invention. In a procedure that dates to 600 BC, the Vedic surgeon Sushruta reconstructed noses from a plot of adjacent facial skin. The "flap" was freed on three sides but remained attached on the fourth in order to maintain a blood supply while the migrant skin grew some vasculature in its new locale. To reach the nose, the flap was swung across and twisted around.

Nasal reconstruction enjoyed a renaissance in mid-fifteenth-century India, this time at the hands of a caste of

---

* When the town of Kirtipur, Nepal, finally fell to Ghurka invaders in 1767, the holdouts' punishment, the story goes, took the form of mass nasal mutilation. All 865 male residents lost their noses, with exceptions made for those who played wind instruments. The last bit puzzled me. Surely one can play a wind instrument without the exterior of their nose. I ran this by the composer and jazz clarinetist David Rothenberg, whom I happened to have recently met. His reply included a link to the paper "Changes in Dento-Facial Morphology Induced by Wind Instruments, in Professional Musicians—A Systematic Review." Playing a wind instrument can apparently be its own form of facial mutilation, bringing on, among other things: receding gums, bone remodeling, reduced facial length, thickened skin, acute pulpitis, lip lesions, overbite, herniated soft palate, "Satchmo syndrome" (rupture of the mouth muscles), recurring tartar, and saliva stagnation.

potters who evidently excelled at sculpting all varieties of malleable material. Here the flaps were lifted from the forehead rather than the cheek. (Long known as the "Indian method," the technique is still sporadically used by plastic surgeons today under the alias "median forehead flap.") By one visitor's account, these were exemplary noses. In *The View of Hindoostan*, Volume II, eighteenth-century Welsh naturalist and travel writer Thomas Pennant marvels at the reconstructed nose of an oxcart driver punished for aiding British colonizers. He deems it "equal to all the uses of its predecessor. . . . It can sneeze smartly, distinguish good from bad smells . . . or [be] well blown without danger of falling into the handkerchief."

However satisfactorily these noses performed or appeared, the rest of the face took a hit. As much skin as it takes to drape a nose, this was the size of the wound. To avoid scarring the cheek or forehead, sixteenth-century Italian surgeons advocated lifting a flap (or "pedicle") from the skin of the inside of the upper arm. The challenge here was that the arm would need to take up temporary—three weeks or so—residence alongside the nose while the flap established itself on the face. You can find period engravings depicting the elaborate network of straps and harnesses used to support the arm and immobilize the graft. The best known of these accompanies a work by the surgeon Gaspare Tagliacozzi, who practiced in Bologna at the inconceivably badly named Hospital of Death. The illustration shows a man with his hand and wrist draped over the top of his head, his face angled to the side as if checking himself for BO.

Tagliacozzi achieved early acclaim as a professor of anatomy at the University of Bologna and wrote a famous medical text on grafting, but his legacy has more or less been reduced to noses. To this day a statue of Tagliacozzi stands in a wall niche at the university's anatomy lecture hall. One foot and an arm extend outside the niche, creating the impression that the man is coming toward us with excitement, to show us what he holds in his hand. It's a nose, presumably from a cadaver. One of two existing oil portraits of Tagliacozzi shows the scholar as a young man seated at a wooden desk, one hand resting on an open book, the other holding a nose between thumb and index finger. Tagliacozzi holds the item with the nostrils facing the viewer, as if to say, *Yes, yes, it is, it's a nose.* Light illuminates what the painter wishes us to focus on: the surgeon's face and hands, his book, the nose. It wasn't until much later that I noticed the severed head, scalp flayed and brain on view, sitting in the gloom on the far side of the desk. It was as though Tagliacozzi had insisted on including it, but the painter was like, *This is too much, Gaspare.* I'm kind of in love with Gaspare Tagliacozzi.

Though the arm flap bears his name, Tagliacozzi apparently wasn't the first to try it. Aina Greig, writing in the *Journal of Craniofacial Surgery*, reveals that arm flap nasal reconstruction was first performed in 1460, by the Bavarian surgeon Heinrich von Pfalzpaint. The surgeon's original manuscripts were lost, and his work has largely been omitted from plastic surgery textbooks. According to Greig, the loss was a substantial one. Four hundred years before the broad acceptance of sterile surgical practices, Pfalzpaint was stressing the importance of keeping clean. He advised the surgeon to bind wounds "with clean white cloths, for if they are not clean, harm results. He should also wash his hands before he treats anyone." Pfalzpaint took operating room hygiene well beyond even what modern medicine asks of its practitioners. "Especially he should guard himself, if he has eaten onions and peas, or slept the previous night with an unclean woman, against breathing into anyone's wound." That Pfalzpaint failed to garner lasting renown perhaps came down to the unscholarly titles he chose for his scholarly works. His treatise on nasal reconstruction is called "HOW TO MAKE A NEW NOSE FOR SOMEONE: WHICH IS OFF ENTIRELY: AND THE DOG HAS EATEN IT."

Even with antiseptic surgical practices, the early success rate for pedicled flap grafting was low. Our army nosemaker Frank Tetamore put it at one in ten, noting that the operations tended to worsen the patient's disfigurement. Many of his clients had been surgically treated before coming to him for a prosthetic nose, in one case having endured ten unsatisfactory operations. And there was the matter of donor site scarring—be it on the face or the arm.

The scarring could be avoided altogether if the flap belonged to someone else. Because of the difficulty of locating a volunteer willing to spend days in bed connected to another human being by a six-inch isthmus of skin, the job largely fell to domesticated animals. The first attempt was undertaken by the esteemed French surgeon Charles Sédillot, in 1868, on behalf of a patient with extensive burns on one hand. In Sédillot's account of the case, the flap was raised from the underbelly of a dog—*un chien Danois*. Offhand, I had no mental image of Danish dog breeds, but given that patient and donor would be sharing a bed for days, I pictured something small and manageable. Not the case. Sédillot had recruited a Great Dane. The effort failed, owing to "*les mouvements excessifs et continuels*" *et* what-the-fuck-did-you-expect "*de l'animal.*"

Perhaps familiar with Sédillot's travails, the next surgeon to attempt the trans-species pedicle graft—E. W. Lee, of County Hospital in Chicago, in 1880—recruited a sheep. It was a wiser choice. Sheep, I've been told by those who raise them, are the chillest of barnyard animals. Unlike pigs and goats, they'll tolerate being made to stand or lie around for protracted periods, as long as they have something nice to eat. There are those who maintain that this is because sheep are not terrifically smart.

Others point to a past of selective breeding aimed at creating animals so docile they calmly submit to being sheared.* The outcome of Lee's work is unknown, as the patient, a young girl with extensive burns to her back, died before the flap was cut loose from the sheep.

At least one other account of a pedicle zoograft exists in the medical literature, this one involving a woman, a pig, and a New York surgeon, all unnamed. The account dates to 1938. The author of the tale, Samuel Lambert, made the point that even were the animal confined, the flap was unlikely to have remained immobile. Many mammals, horses most notably but also pigs, possess the ability to voluntarily twitch the panniculus carnosus, a thin, broad muscle just below the skin, when they wish to dislodge an irritant—typically flies but in this case a human. Lambert's account also makes reference to resistance on the part of ward nurses tasked with clearing livestock excrement from a hospital room multiple times a day. And that, more or less, marked the end of the barnyard pedicle flap.

But not the end of the barnyard skin donor. Nineteenth-century surgeons were discovering that a skin graft needn't remain attached to its homeland, that healing would take place when the flap was cut completely free from its original owner (a "free flap," as opposed to a pedicle flap). To find donors, surgeons would turn first to family and friends, but, as Scottish

---

* Either way, it helps explain the enduring popularity of sheep—lambs, especially—in seventeenth- to nineteenth-century blood transfusion experiments. Medical journals of the day featured woodcut illustrations of lambs lying, seemingly contentedly, alongside bedridden patients, the pair connected neck-to-arm by a tube. I had wondered whether belief in some biblically derived Lamb-of-God purity explained the choice, but no article mentions this. Regardless of the reason, the practice was ill-advised—as well as impractical. As one practitioner wrote in an 1876 issue of the *Medical Examiner*, "will a lamb be at our disposal just when we want it in such cases of emergency?"

surgeon Alexander Miles pointed out, "such self-sacrificing mortals are few in number." An 1898 news piece in the *Brooklyn Daily Eagle* laid out some numbers: 817 pieces of human skin were needed to cover the burns of a young woman whose clothing caught fire when vapors from the gasoline she was using to clean a pair of gloves ignited. Once the physicians had exhausted the acreage and forbearance of family and friends, they resorted to the skin of strangers. An ad placed in a local newspaper offered payment to people willing to part with a snippet of skin. Five hundred individuals responded—though about a fifth of them were "men who understood the advertisement to read that men were wanted for tree grafting and had come ready to go to work." Even with the added volunteers, 137 pieces of chicken skin were required to make up the deficit.

The chicken was briefly the donor of choice among nineteenth-century zoografters. The rationale was one of surgical practicality. The skin under the wing is loose, with no fat adhering—ready to go with very little prep. France's Paul Redard insisted upon young chickens. Youthfulness, rather than species, was thought the more important criterion in zoografting. "I have always used young animals," wrote Alexander Miles in "Observations on the Grafting of Skin Taken from the Lower Animals," ". . . in the belief . . . that the comparatively great developmental power of young tissues will enable the new covering, once it has adhered, quickly to spread over the raw area." Miles's contemporaries shared his belief. M. E. Van Meter was a puppy guy. G. F. Cadogan-Masterman preferred young rabbits. Thomas Raven tried a piglet. Miles summarized his own flensings in a handy table: eight kittens, five puppies, and two four-day-old rabbits. Who's the "lower animal" now?

To be fair, Miles and his fellow puppy and kitten flayers operated a century before the widespread establishment of spaying and neutering programs. Strays from unwanted litters of dogs and cats were legion and viewed with little sentimentality. A similar combination of availability and public indifference to the creatures' welfare catapulted frogs into the skin grafting spotlight. Writing in the *British Medical Journal* during World War I, Captain H. W. M. Kendall explains how he came to experiment with frog skin as a covering for the shrapnel, bullet, and bomb wounds he found himself treating. Kendall was stationed in France, where, as he put it, "demand and supply of material are both abundant." I imagined him striking a deal with local restaurateurs: *I can take those frog torsos off your hands, Chef.* But Kendall, like everyone else in France, preferred the legs. The inner thighs, he noted, have very loose skin, easy to use. Just like chicken.

Frogskin grafting caught on, spreading in time to the United States, where it made headlines not only in medical journals but in daily newspapers. Coverage ran the gamut, from the jauntily alliterative "Repaired with Flesh from Friends and Frogs" to the frankly tasteless "Frog Skin Used in Effort to Keep Texas Woman from Croaking." Only the *Times-Mail* of Bedford, Indiana, managed to maintain its composure: "Skin of Thirty Frogs Is Grafted on Woman." No big deal. Move along. (Though composure went out the window a few column inches down the page, with the headline "Cow Is Ventriloquist: Able to Throw her 'Moo' All Around the Lot.")

Though practitioners at the time wouldn't have been aware of it, frogskin is apparently a rich pharmacopeia. According to James Crowe Jr., author of the *Immunity* paper "Treating Flu with Skin of Frog," peptides isolated from skin of frog (I'm retaining the Shakespearean phrasing) possess not just antiviral but also antibacterial and immune system–modulating properties, both of which could, who knows, have benefited graft recipients.

The Frankenstein element of zoografting proved hard to resist. The *Oakland Enquirer* from June 26, 1911, relates the story of five-year-old Thomas Reardon, recovering from severe burns and soon to be "the only person in St. Louis who can exhibit a frog skin leg." While some surgeons reported that zoografts sloughed off and understood them to be temporary coverings, others suggested that patients would go through life with functioning plots of the living skin of another species.

William Allen, writing in the medical journal *The Lancet* in 1884, presents an unusual take on the matter. He posited, bewilderingly, that healing was achieved through a kind of dermal sexual reproduction. "If the sexual theory accounts for the process, the skin that grows after the application of the frog grafts must be of the nature of a new breed, a cross between human and frog epidermal elements." Frankenskin.

I have questions, many questions. The cells of zoografts—or xenografts, as they are called today—are surfaced with foreign proteins. Why wouldn't the patient's immune system reject them? Why would they "take" at all? And if they're rejected, what was the point? Whose skin do plastic and reconstructive surgeons rely on today?

For answers, I'm inflicting myself on the generous professionals of Massachusetts General Hospital, home of the Sumner M. Redstone Burn Center. The facility is one of the nation's largest and best-known burn units, the hospital having gained prominence in the aftermath of the 1942 fire at the Boston nightclub Cocoanut Grove, which killed 492 people. Thirty-nine burn patients were admitted to Mass General that night. The hospital archivist sent me an illustrated *Annals of Surgery* paper summarizing their treatment. Nine had required skin grafts, often on the backs of the hands, because people had covered their faces. Though the heat was so intense as to burn off the tip of one woman's index finger, the nail polish on her remaining fingers survived intact. One's eye fixes on details like this, because everything else is too horrible. It's no mystery that religion's showrunners chose fire as the setting for Hell.

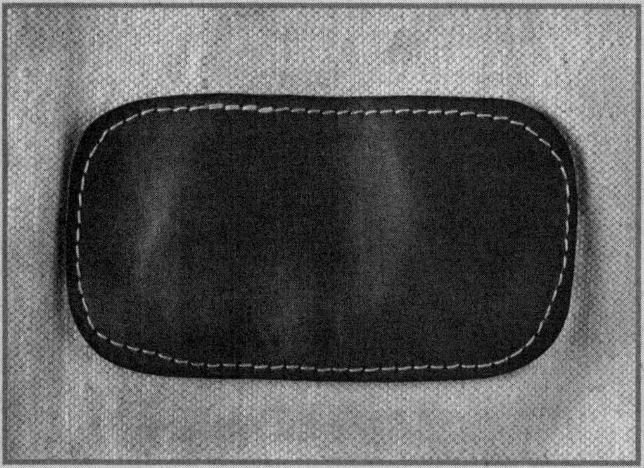

2
Gimme Some Skin

Replacing the Human Exterior

You've seen the textbook illustrations: the big cube of human skin with its confectionary layers—the pink frosting of the epidermis, the paler dermis, the orange fat below. Big and pretty as a petit four. In reality, we are as drably thin-skinned as chickens. The epidermis is maybe thirty cells deep. The dermis, a few millimeters at its thickest. Earlier today, in the Massachusetts General Hospital burn unit, I watched an intern named Seamus plane a skin graft a third of a millimeter thick. You could have folded it and mailed it with a first-class stamp.

A skin graft is thin to help it survive in its new location. Until capillaries from the wound bed begin to grow in, the cells of the graft will be nourished by the plasma they sit in. The term for this is osmotic imbibition: drinking through your membrane. If the graft were much thicker, the cells on the inside of it would starve. Taking a very thin graft also speeds the healing of the new wound, the one Seamus had just created.

The graft I saw was planed from skin on the patient's own thigh. This is autografting. You are your own skin donor. It is the ideal skin grafting scenario. Anyone else's skin—spouse, cadaver, frog—will eventually be rejected by the immune system. Anyone else's is a temporary covering. It may be called a graft, but more accurately, it's a kind of dressing—a biodressing. This has been explained to me by Jeremy Goverman, a dark-haired, gifted, unwaveringly genial plastic surgeon at Mass General's Sumner M. Redstone Burn Center.

For the past hour, Goverman has been answering my questions while simultaneously toggling between two computer monitors, updating patient charts and billing forms. Calls come in from the OR downstairs, and texts—*ting!*—land every half minute or so. A sign on Goverman's desktop reads: Good Enough. It's there as a reminder. Goverman is driven, a self-described perfectionist. He has struggled with a health care system that doesn't prioritize—or leave time for—doing things perfectly. "Acceptance" is written on a yellow sticky note on the frame of one of the monitor screens. You may recognize it as a tenet of AA. Between 13 and 25 percent of surgeons grapple with substance abuse at some point in their careers.

Currently testing Goverman's newfound equanimity: a visitor with no medical background and a long list of questions, many of which, it will become clear upon transcribing her recordings, she is asking twice.

"Wait, so why would you use anything other than an autograft?"

Goverman swivels from his computer to answer. Supply problems, in short. With a major burn, surgeons quickly run out of graftable (unburned) skin. "Sometimes we have to use the soles of the feet, the scalp, even the scrotum," Goverman says. Areas that have been harvested can be reharvested after they've healed, but that can take two or three weeks. In the meantime, an allograft (skin from another human, typically a deceased tissue donor) or a xenograft (skin from another species) protects the wound. It also prevents some of the fluid loss. Deep burns destroy the closed-loop integrity of the circulatory system; leaking capillaries (among other things) send blood pressure dangerously low. Temporary grafts also make changing bandages less painful for patients, and they help keep them warm. Because they've literally lost their coat. Burn survivors can develop hypothermia in a 70-degree room. ICUs and ORs for survivors of major burns are often kept at 90 or even 105 degrees, nurses and surgeons sweating under their scrubs and sterile gowns.

Another question for Goverman: If skin grafts taken from other bodies are just a temporary dressing, why then did the frog and dog and rabbit skin grafters of nineteenth-century medical journals (and of the previous chapter) write that these grafts would "take"? Because they did. "It becomes your skin for a few days," Goverman says. One of the things that happens with a big burn is that the immune system is suppressed. The lowered vigilance means a piece of foreign tissue is not flagged as dangerous. So the body follows normal skin protocols, extending the welcome handshake of capillary growth. For a matter of days, people were in fact part frog or puppy or rabbit.

An allograft lingers longer under the radar, because skin from a fellow *Homo sapiens* is a closer match than skin of another species. It likely buys you weeks, not days. Whatever is used, person or beast, a graft also helps prevent infection. The immune system's temporary slowdown makes this critical in the early days after a serious burn. When someone dies after a serious burn, it's often because of sepsis: a localized infection has gone systemic.

Eventually an allograft for a bad burn will be replaced with a permanent graft of the patient's own skin. The surgeon can wait for the allograft to be rejected and slough off, or the surgeon can, as I believe I phrased it in a question to Goverman, rip it off.

"Ripping it off is actually nice because it freshens the wound bed." Makes it bleed, he means. Blood is good. Blood heals. Blood feeds the new graft. "You want a nice vascular wound bed." It's unusual to encounter *freshen*, *nice*, and *bed* outside of, say, a hotel website. That's surgery for you.

Goverman takes another call. It's the OR again, letting him know his patient is prepped and ready. We walk to the elevator, Goverman slightly ahead of me, *ting*ing as he goes. Because he's so good-natured, smiling and joking with hospital staff, the *ting*s have begun to seem, to me, like a feature of his personality, a sort of aural sparkle.

The patient is a man, youngish, already sedated on the operating table. Goverman wheels over on a stool and takes one of the man's hands in his own. *That is really touching*, I think. Then he begins wiping the back of the man's hand with a blue surgical cloth—briskly, the motions of a parent cleaning something gooey from a toddler's hands. He's wiping away dead skin. This helps him assess the damage, which can be hard to do at first. Second-degree burns can progress to third-degree in the early aftermath of a serious burn. And this one was serious. The man had been trying to fix an oil furnace, someone is saying. He touched two wires together, causing a spark and an explosion.

"Let's not talk about that," Goverman says quietly. Survivors' stories, he feels, are their own to tell. I knew this, yet when I accompanied him on rounds, earlier, I was unable to stop myself. *What happened, what happened, what happened.* I can tell you this. Oil and grease are trouble. Boiling water, because it runs off, has limited time on the skin. Something oily will stick, and burn longer, as will anything viscous. There is a mac-and-cheese burn on the floor above us.

Goverman moves to the man's wrist, using a tool now, scraping with short strokes, like a lotto player getting busy with the edge of a coin. Excision, or debridement, removes dead cells and bacteria and other bio-garbage that can interfere with healing. The deepest burns are on the man's calves. He must have been wearing shorts. The skin here is about to be allografted. Goverman prepares the site with a longish blade called the Goulian (pronounced, with some aptness, *ghoul-ian*). Strips of flesh are pared with a quick, truncated back-and-forth motion. *Where have I seen that*, I think. A moment later, it pops to mind: shawarma shop.

A nurse arrives with the allografts, four sheets, packaged and sealed. They'd been in a freezer until this morning, when someone thawed them in a bowl of room-temperature saline. (DO NOT MICROWAVE, the package insert warns.) The skin is pre-meshed in a diamond pattern, like an expandable baby gate. The mesh allows the surgeon to stretch the graft as needed, to cover more burn. (The meshing also allows fluid to drain.) An autograft removed on the spot, in the OR, would have to be meshed by hand. On a table against one wall is a mesher, a miniature version of what a chef would use to cut sheets of fresh pasta. Meanwhile the dermatome, the instrument used to remove an autograft, is a sort of medical-grade cheese slicer. Kitchens and operating rooms—the ever-present, mildly disconcerting overlap.

Using forceps, Goverman spreads the allografts and staples them in place. They give the appearance of thick, pale fishnet tights. The hospital recently switched to a less expensive supplier, and Goverman doesn't like it. "It's slimy, and the mesh pattern is variable." Even with cadavers, you get what you pay for.

The newly placed grafts are covered with an antibacterial foam dressing called Mepilex and then mummy-wrapped in white gauze and finished with a winding of elastic bandage. Goverman calls out the square centimeters, and a nurse makes a note, for billing.

Goverman used cadaver skin today, but he has, in the past, used a pig skin product, and he recently went through an Icelandic cod phase. The cod people had invited a number of what are called key opinion leaders to come see the process and learn about the product. Junkets are one way that companies get products into hospitals and surgeons' hands. Goverman makes no excuses. "I was like, *I wanna go to Iceland!*"

And? Is the cod product better than a traditional cadaver allograft? We've gone back up to his office, so he can dictate a surgery summary while the work is fresh in his mind. "They have some studies, some basic science—put it on a wound and find some markers that go up or down. *Look, there's less inflammation,* it's modulating this or that. Most of my colleagues think it's great." He's not sure. "I will say, when you put it on the wound and you wait a week, the wound bed looks pretty good." He's typing again, like he's finished with what he's got to say about cod skin. Then stops. "But when you do nothing"—can't help himself—"and you wait a week, the wound bed looks pretty good."

Here's an advantage: The hospital can bill for a skin substitute. "Mepilex will get you ten dollars," Goverman is saying, "and fish will get you a thousand." There are yet pricier options. Companies tout the advantages of grafts made from human embryonic membranes, placentas, dehydrated umbilical cells, foreskin* cells seeded onto cow collagen. They claim that they're rich in generative factors and less antagonistic to a patient's immune system. In fall of 2024, the *New York Times* ran a hyperventilatory piece entitled "Her Face was Unrecognizable After an Explosion. A Placenta Restored It." I sent the link to Goverman. Her face looks great, he replied, because the head has robust blood flow; thus a second-degree burn to the face heals "spectacularly well" on its own, with nothing fancier than an antibiotic ointment.

There are seventy skin substitutes on the market right now, Goverman says. "There's big, big money in this stuff."† And is

---

* The use of foreskins as allografts dates at least as far back as the 1930s, when physician Frank Ashley counseled *Annals of Surgery* readers to drop by hospital maternity wards, where "one may obtain all the foreskins necessary." The discarded prepuces could then, he advised, be stored in a jar "in a refrigerator" or "imbedded in ice cubes"—either option with the potential to enliven, or derail, impromptu baby arrival celebrations on the maternity ward.

† For the budget-minded: boiled potato peels. Yen-Fan Chin, writing in the Taiwan journal *Hu Li Za Zhi*, refers to clinical trials showing that the boiled peels fared as well as two commercial wound dressings in terms of the pain experienced during dressing changes. Scoff not! A PubMed search of "bioactive" and "potato peels" turns up fourteen articles on the medical virtues (antioxidant, antibacterial, anticarcinogenic, anti-inflammatory) of compounds found in the peels of the humble tuber.

there truth to the claims—that products made with cells from the earliest stages of development promote faster healing? "We really don't know," Goverman says. Nor, he suspects, does the U.S. Food and Drug Administration (FDA). Biodressings are reviewed by the Division of Dermatology and Dentistry, though neither profession treats survivors of serious burns.

"What we do know," Goverman adds, "is that despite all these products, wounds and open burns are still a huge burden and we're still using, ultimately, the patient's own skin." (In 1936, a physician named Jacob Sarnoff combined both approaches, circumcising a young boy and then grafting his foreskin directly onto his finger, which had been "degloved" when his ring caught on a fence spike.)

Most autografts take the form of dermatomed strips immediately applied in the OR, though patients' skin can also be cultured off-site, from a biopsy. There's an aerosol autografting product, Spray-On Skin, that's used alongside widely meshed autografts, to speed regrowth in the spaces of the mesh. Goverman often uses CEA, cultured epithelial autograft, wherein the patient's cells are grown into a very thin—two or three cell layers—sheet of epidermis.

Even better would be a thick sheet—a full-thickness cultured autograft with dermis as well as epidermis, to use with severe burns. The Swiss company CUTISS is about to enter Phase 3 European clinical trials of one such item, denovoSkin. Early this morning, during rounds, Goverman checked in on a young boy who'd recently been admitted with burns on 90 percent of his body. The room was sweltering. Bandages covered everything that had not been burned away or amputated. Through an FDA compassionate use exemption, the child will become the first American burn survivor to receive a denovoSkin graft. After clearing customs, a biopsy of the boy's skin will fly to Zurich in a refrigerated box. There it will be separated into dermal and epidermal components that will be cultured and then seeded onto a sponge-like collagen scaffold. About a month later, a piece of skin the size of a large elbow patch will fly back to Boston with its Swiss escort, a surgeon who'll help put it in place. If all goes well, more patches will follow. Goverman hopes that will be the case, for the sake of the boy, but also for CUTISS. In medicine, the products with the most meaningful impact are rarely the biggest moneymakers. More than $70 million has gone into the development and testing of denovoSkin, and the market is comparatively small. As opposed to, say, bio-bandages made from fish farm waste.

The fish skin trend started in Brazil with tilapia skin. "It worked well," Goverman allows, "but they were using it on second-degree burns, burns that would heal pretty well on their own."

Second-degree versus third-degree is an important distinction. Burn degrees are like Richter earthquake numbers—one step up doesn't sound like much, but in terms of the damage wrought, it's substantial. With a second-degree burn, skin regenerates from below, from cells in the dermis. You begin to see little islands of new growth, called granulations. (Cases of overgrowth are known, lyrically, as "exuberant granulations.") This process can't happen with a third-degree burn, because the burn has destroyed those generative cells. The body resorts to a different strategy. It tries to close the wound by contracture. The skin around the perimeter of the burn pulls together, like the waistband on a pair of drawstring pants. This is the real nightmare of recovery from a major burn. As it pulls inward, the skin around the burn pulls on the rest of the skin. The contracture leaves the patient disfigured and can hinder their movement.

Goverman shows me examples, photographs, of how contracture can distort a body. A chin pulled down into a collarbone. An elbow permanently bent. Eyes held open, unable to blink. He clicks to a slide of a woman, Diana Tenney, who suffered third-degree burns on 90 percent of her body. In the image, a long open slash traverses her midsection. This is part of her recovery—the first stage of a contracture release. The cut relieves the tension of the pulling. The skin on either side recedes, and the resulting gash is filled in with a full-thickness autograft. Because the donor graft site would likely scar, the skin is taken from a less conspicuous place. That place, in turn, is grafted with a split-thickness graft from somewhere else on the body. Robbing Peter to pay Paul to pay Lou. And then coming back a month later and robbing them all again. Some places on Diana's exterior were reused for grafts five, even six, times. Her recovery spanned seven years and more than twenty-five trips to the OR. Diana and her husband, Jerry Laperriere, are joining us for dinner.

She has told the story a hundred times. To strangers beside her in checkout lines. To curious children. To the Walmart shopper who looked up from her cart and screamed, "OH MY GOD, OH MY GOD, WHAT HAPPENED TO YOU?"

One more time, she is telling it. We're at the restaurant of the Liberty, the hotel that is semi-attached to Mass General and semi-fancy (lobster pizza). "I was up in a tree, sawing off branches to get more sun in my backyard. Jerry came out and offered to finish up, so I came down and started stuffing the branches in the chiminea. They were green, so they didn't light. I went through a whole can of lighter fluid." She spreads her napkin in her lap. "Well, it still wasn't lit, so Jerry got the gas can out of the shed." The fumes combusted as Diana was walking by. "And my skin was soaked with suntan oil. I was like a wick! I was trying to look for a place to 'stop, drop, and roll,' but every time I looked down I saw the ground on fire." In fact, it was herself she was seeing on fire, not the grass. The restaurant is loud, but the quiet at this table somehow overrides it. "I ran a circuit around the house. We estimate I was on fire for about seven minutes."

"We didn't have the hoses out," Jerry interjects.

"It was early March," Diana adds. "A beautiful day." That last detail is a wallop. To have gone, in a second, from *beautiful spring day with someone you love* to *I am on fire*. "Fortunately, our neighbor, who washes his jeep every single day of the year, had seen the fireball and ran over with his hose. He didn't notice me at first. He was hosing the awning and the fence. Jerry saw him with the hose and asked him to put me out. Which he did. And I just kind of sank down by the side of the house." There was no pain, she adds, because the fire had burned away her nerve endings.

The next six months were a blur. "They put you in an induced coma for a month, and for another four or five months after that, they've got you on so many meds, you don't really remember much." Fourteen, at one point. "My life was surgery, recovery, surgery, recovery." An infection. More surgery. More recovery. Goverman did eight surgeries just to release the contractures on her neck. One of them hit a salivary gland, Goverman recalls. "So there was this point where she would see food . . ."

"I'd be out to eat and my neck would start dripping." She and Goverman are laughing. Jerry isn't. I don't know him, so it's hard to say whether he's more somber by nature or whether the ordeal has stolen some lightness. I would guess the latter. For the first few months, the fear and anxiety were mostly his. He would see patients come into the ICU with less serious burns than Diana's and not make it out. "I'd say to the doctors, 'How's she doing?' They'd say, 'She's very sick.' I'm like, 'Well, I know *that*.' I'm almost positive most of them didn't think she was gonna make it." With reason: Only around 5 percent of patients in Diana's situation—a fifty-four-year-old with burns on 90 percent of her body—survive.

There are ways to preserve hope without lying or discounting the seriousness of a situation, and Goverman was adept at that. Others, less so. "One of them said to me, 'Maybe next time she gets an infection, we should think about how we treat it,'" Jerry says. "He meant *if* we treat it. As in 'comfort measures.'"

For Diana, the pain, emotional and physical both, began in burn rehab: the daily lineup of physical therapy, occupational therapy, speech therapy. Worse than all of that: the first time she saw herself in a mirror. At Mass General the staff were careful to cover reflective surfaces, but a physical therapist took her to a gym that had a full-length mirror. "It wasn't supposed to happen until much later," she recalls. "It was a major major major major trauma." She looks over at Jerry. "I'm thinking, he's going to leave me. I couldn't understand why he was still there, unless it was out of guilt."

Jerry sets down his fork. "I told her, 'It's just skin. I don't love you for your physical appearance. I love you because you're you.' And she didn't say anything. And I said, 'Are you still you?' She said, 'I'm still Diana.' I said, 'Then I still love you.'"

About her physical appearance. You can tell Diana was seriously burned. You can detect the patchwork of grafts. But when you look at her, mostly what you notice is that she's beautiful. It's a physical beauty, a testament to her features and to Goverman's skills as a plastic surgeon, but it's also a presence. She seems content and self-assured. I noticed that as we were seated at our table, she chose the chair facing the room and the other diners. She laughs easily and meets people's gaze. And she's wearing a sleeveless dress—not to make a point, but because why wouldn't she? And here I am in my long-sleeve shirt on a hot summer day, because I'm sixty-four and I have arm hang.

Like Goverman, Diana has been through addiction rehab. Like him, she credits it with her ability to get through a very bad time in her life. The two see similarities in recovery from burns and recovery from addiction, in the mindset and grace it instills—acceptance and gratitude and a host of other things you can't get from spray-on cells and fish skin.

Before I left, I asked Goverman about the future of burn care. He told me that he was in the midst of a trial of pig skin that had been genetically altered to be more like human skin—or, more precisely, less like pig skin. The aim being to create a xenograft that functions more like an autograft. A couple of companies are working on this sort of thing.

If all goes well with this line of research, gene-edited pigs will eventually be donating more than just skin. The science is moving from xenografting to xenotransplantation. "The ultimate goal would be, you'd have your own personal pig," Goverman said. In other words, a pig edited to match your own genetics: skin, kidneys, heart—ready to use, like a car kept for parts.

As I write this, gene-edited pig hearts, kidneys, and livers have been transplanted into patients with failing organs and no hope of other treatment. The surgeries, especially the first, garnered a glut of media coverage on the operation and the patient. I was interested in the pigs, some of which are raised in Iowa. The juxtaposition of midwestern pig farms and cutting-edge biomedical research appealed to me. I took a Zoom meeting with the CEO of eGenesis, the company that supplied the first kidney, to talk about spending a day with their "chief pig engineer." It went, I thought, very well. And then they ghosted me. Their competitor United Therapeutics, the firm that created the transplanted pig hearts, also nixed a visit.

I turned next to Shaoping Deng, coauthor of a paper in the journal *Xenotransplantation* entitled "Xenotransplantation in China: Present Status." At the time of the paper's publication, 2019, Deng worked at the Sichuan Academy of Transplant Science, in Chengdu, where he had been for the last thirty years, working on xenotransplantation. One of Deng's coauthors on the paper, Yi Wang, replied to my email: "If you are interested in the progress we have, you can visit our institute and the animal institution." Huzzah! Exuberant granulations! We are off to Chengdu.

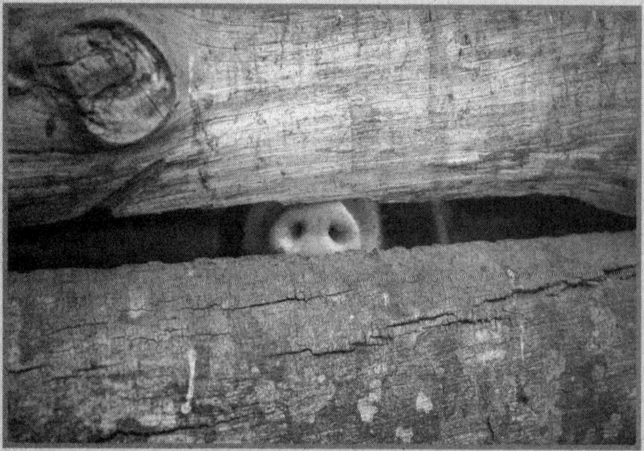

3

Mixed Meats

*Humans with Pig Organs,
and Pigs with Human Organs*

By one metric of the economic rivalry between China and the United States, China is far ahead, and that is pigs. The data analytics firm Dun & Bradstreet reckons that the U.S. has around 7,000 pig farms. China has more than 150,000. Half the planet's pigs are there. The province in which I find myself, Sichuan, has 16,000 pig farms, second only to Shandong. Sichuan is known for pandas, but IMHO, pig farms steal the show. Piggeries 26 stories high with elevators that lift 40 tons. Central manure processors. Pig facial recognition!

Based on my time in Sichuan's capital, Chengdu, China feels far ahead, period. The AC in my hotel room works by voice command. On a 3D billboard on Chunxi Road, I watched a pair of pandas that, were they not the size of Macy's balloons, I might have believed were real. I rode across town in an electric car whose owner, rather than stop to recharge it, backed it into a stall for a quick robotic battery swap.

Here's one place China lags: organ donation. Amid the fast-forward juggernaut of AI and QR and 3D, Confucian beliefs hold strong. "In Confucianism . . . the physical body is considered as a gift from one's parents," write Shaoping Deng, Yi Wang, and their *Xenotransplantation* coauthors. Yi Wang, like Deng, has spent years working on xenotransplantation at the Sichuan Academy of Medical Sciences, of which the Academy of Transplant Science is a part. "Thus, any damage inflicted upon the human body—even in death—is thought to disrespect ancestors." This belief system, their paper goes on to say, explains the "reluctance" to allow one's organs to be removed for donation. Yi Wang puts it more bluntly in person: "In China, we are unwilling to donate."

And so, for decades, death row prisoners have been China's source of organs for transplant. In 2014, Jiefu Huang—a transplant surgeon who'd recently served as a deputy minister of health—announced that the practice was about to stop. In fact, the change has amounted mostly to semantics. "Death-row prisoners," Huang told the *Beijing Times*, "are also citizens and have the right to donate organs. Once the organs from willing death-row prisoners are enrolled into our unified allocation system, they are then counted as voluntary donation from citizens; the so-called donation from death row prisoners doesn't exist any longer." Crafty!

Deng told me six thousand Chinese are registered in the nation's organ allocation system. No matter who they are, it's an infinitesimal drop in a very large (1.4 billion people) bucket. Even if China had a robust pool of potential voluntary donors, need would still swamp supply, as very few of them will end up in a position to donate. In China, death is still defined by cessation of the heartbeat, rather than by brain death, which is the legal criterion in the U.S. What is sometimes here called a beating heart cadaver—a brain-dead (that is, legally dead) individual whose heart and other organs are being oxygenated via life support equipment—would not, in China, be considered a cadaver at all. To remove organs for transplantation would constitute not only disrespect but murder. Thus the only potential organ donors are terminally ill patients on life support whose families agree to have that support shut off inside an operating room set up for organ recovery. And of course, "willing" death row citizens.

A potential solution to China's organ shortage combines two of its strengths: technology and pigs.

Gene-editing techniques like CRISPR make it possible to tweak a pig's genome such that its organs, if transplanted into a human, will seem less foreign—less immunologically triggering. Unaltered, a pig's cells bear a surface protein called alpha-gal, which is a blaring, flashing red alert of mammalian otherness. The human immune system responds with "hyperacute rejection." Within minutes, a pig kidney transplanted to a human will begin to turn dark red and then black as elements of the patient's immune system attack and destroy it. In 2001, the U.S. biotech firm Revivicor—later acquired by United Therapeutics—announced it had created pigs with the gene for alpha-gal "knocked out," to use the insider phrasing. Researchers in China have followed suit, creating their own pigs whose organs, if transplanted, won't be destroyed by hyperacute rejection.

Longer-term rejection remains an issue. Thus a patient who receives a "gal-knockout" organ—a UHeart, say, or a UKidney, as the "Unitherians" of United Therapeutics have trade-marked their products—would need to be on a regimen of immunosuppressive drugs similar to what they would need after receiving a human organ. As the technology stands now, a pig organ is only an advantage if the other option is no organ at all. For the roughly six thousand Americans and Lord knows how many Chinese who die each year waiting for a human organ to become available, that represents a weighty advantage.

As I write this, two Americans have received gal-knockout pig hearts. Because the FDA has not yet approved UHeart for transplant, the surgeries were done under the agency's compassionate use exemption; both patients would soon have died without the procedure and were too ill to qualify for a human organ. Both survived the transplant but died within a couple months, having contracted a pig virus. Complicating matters for one patient: the heart began to expand beyond the rigid confines of its pericardial lodgings. Researchers had used the heart of a one-year-old pig, gene-edited to halt the organ's growth, but it began to enlarge for other reasons. That, combined with swelling and inflammation from the virus, ruined the heart.*

On the kidney front, an eGenesis EGEN-2784 gene-edited pig kidney landed in a patient at Massachusetts General Hospital in March 2024, with a second transplant following in April. Both patients survived around seven weeks. That sounds, on the face of it, discouraging, but is less so if one thinks of a pig organ not as a lasting fix but as a bridge to such a fix. A pig organ could potentially buy a patient enough time and health to become eligible for a human organ.

The media buzz surrounding the first pig-to-live-human xenotransplant made the work seem new, but in fact, xenotransplantation research has been underway for decades. Shaoping Deng has been at it thirty-three years. I'm sitting next to him at a teacher appreciation dinner being held for him and fellow xenotransplantation researcher Yi Wang. I certainly appreciate Yi Wang. Among her many other duties as a professor and researcher, Yi answers Shaoping Deng's email. It was she who replied to me and invited me to visit, who booked me a room through the hotel's Chinese-language reservation system (and conversed with the Chinese-speaking air conditioner).

The dinner is taking place at a popular banquet-style restaurant not far from the Academy of Medical Sciences. Six graduate students sit around a table while a lazy Susan drifts, barge-like, delivering platters. Deng is dressed, as yesterday, in his uniform of panda colors: black trousers and white dress shirt. He is sixty but shows no wrinkles, in his face, his shirt, his enthusiasm for the future of xenotransplantation.

---

* As well as, for me, the happy ending of *The Grinch Who Stole Christmas*, wherein his "small heart grew three sizes that day."

Setting aside the ethics of building short-lived pigs to lengthen human lives, the therapeutic promise of pig organs is tremendous. You can't edit the organs of a live human donor, but with pigs, one can tinker as long as the funding holds. Chinese and U.S. researchers create pigs with up to seventy edits: pig genes knocked out and human genes knocked in. "But in the future?" Deng says. "We could modify far more." I'm listening, but I'm also watching for the Sichuan braised whole fish, because I missed it on the first rotation.

Could the technology of xenotransplantation bring us to the point, conceivably, where a pig's organ could be a closer match for a patient than another human's organ? "Absolutely right! Good point." Deng's speaking style is decisive, exclamatory, delighted. (When I said I'd run out of questions yesterday: "Good! Save my time!") "Through the genetic manipulation, a porcine organ could be *better* than a human organ!"

Hundreds of proteins differentiate a pig's organs from a human's. Would geneticists try to edit everything that doesn't match? That seems crazy.

"We cannot make them identical," Deng says. "But! We can change the organ so it will produce the immunosuppressive protein. So when you transfer it to the human body, it can secrete the immunosuppressive protein." Localized immunosuppression, he's talking about. The organ protects itself. So its owner would no longer require a lifetime of systemic immunosuppression—the toxic drugs that leave transplant patients vulnerable to infections and other long-term side effects.

The fish comes to a stop in front of me. In my eagerness, I forget to switch to my "public chopsticks," the set meant for serving oneself. A grad student sees this and looks away. Can someone knock in some manners?

How many years, I ask Deng, will it be before organs like UHeart and UKidney are routinely put to use in patients and the procedures covered by insurance? Optimistically, he says, five years. "Conservatively? Ten." It may be a long U-Haul. Deng extracts a skinny cigarette from a pack. A student quickly leans in to light it. It will happen first in the United States, he says, followed by Germany. "China will be third."

And then it will catch up and overtake the world. One day, I say to Deng, China will have 26-story organ farms.

"Yes! You are right." He's familiar with the famous pig highrise. "That's the future!"

For now, there is a single-story farm. The pig facility of the gene-editing company ClonOrgan is in Neijiang, a two-hour drive from Chengdu. I'm visiting it tomorrow with ClonOrgan's founder, Dengke Pan. Pan has created multiple varieties of transgenic pigs and, in 2005, produced the first cloned pig in China. Yi will come along on the road trip to help with translation. Pan speaks English moderately well, but Yi is fluent. She earned a graduate degree in Ohio. When I asked her how she survived without the spicy glories of Sichuan cooking, she replied, "Wendy's chili." I made a mental note to try it sometime.

It's generous of Yi to take a day away. She has a crushing schedule of teaching on top of the demands of research and parenting (and Shaoping Deng's email). Academia is a life she does not wish for her son. She sometimes urges him to skip classes, to go swimming instead. She told me she wants him to be a car mechanic. I couldn't tell if she was joking. She often is. She is vibrant and resilient in a way I could never, under the circumstances, manage.

At the end of a tour of ClonOrgan's Chengdu labs yesterday, Dengke Pan presented me with a gift. It was a pair of plastic pigs, pink with a black patch around their tails and another on their heads, the top half only, resembling a Batman mask. "Just like the ClonOrgan miniature pig," Pan said. He meant that the toys shared the pigs' unique coloration, and rationally I knew this, but for one brief, delirious moment, I imagined that he had created pigs the size of hamsters. Pan maybe sensed this. He looked at the pigs in my hands and smiled. "*Miniature* miniature pigs."

The original miniature pigs were bred in the United States in the 1950s, not as pets, but with the aim of creating an ideal animal for surgical and medical research: a creature bred "small enough to be easily handled" without changing the size or function of the organs, because they are a decent match for our own. Monkeys and apes are of course closest to us—genetically, anatomically, physiologically—but that closeness creates ethical quandaries. And diseases are more readily shared between fellow primates—a key concern for xenotransplantation. Also working in pigs' favor (or disservice): they reproduce quickly and abundantly, are cheap to feed and house, and 60 percent of the world is already routinely raising and slaughtering them.

In centuries past, dogs and cats were the species most commonly used for surgical experimentation. Here is Russian physiologist Ivan Pavlov on why he preferred dogs to pigs: "As soon as a swine was lifted onto a stand, it squealed at the top of its voice, and all work in the laboratory was impossible. . . . All pigs are hysterical." (The collective noun for wild pigs is a "sounder.") And big. They're a big animal to have around the lab.

A formal "miniature swine development" project got underway in 1949, a collaboration between two Minnesota powerhouses, the Mayo Foundation (research arm of the Mayo Clinic) and the Hormel Institute (research arm of pork). Mayo had the need, and Hormel had the pigs. It wouldn't be hard. Hormel had been breeding bigger pigs—more pork!—and the same methods could be undertaken, as one historian of the program put it, "in reverse direction."

The National Heart Institute—then a branch of the National Institutes of Health (NIH)—was especially excited. Pigs bred for pork and raised in confinement were fitting subjects for the study of heart disease in humans: They don't exercise, they've got large stores of fat, and they get atherosclerosis at a young age. (And they eat garbage.) "Breeding for market purposes has produced an animal which from birth develops as a counterpart (or even a caricature) of the obese and sedentary human," wrote the author of the paper "Swine in Comparative Cardiovascular Research." All that was needed, he added, was a smaller "edition."

Hormel's miniature swine project spanned fifteen years, encompassing not just efforts to breed them smaller but far-ranging investigations into pig physiology as it compares to ours. The papers are collected in three thick volumes. "Some Comparative Aspects of Porcine Renal Function." "Microscopic Anatomy of the Skin in Swine." "The Miniature Pig in Dental Research." (Pigs, yes, wearing braces.) Because so much was now known, the pig became the go-to mammal for medical-surgical research and remains so today for trials of gene editing and xenotransplantation.

How do we know the organs of some other smallish domesticated mammal aren't a closer match? A goat's heart, say, or llama kidneys? I posed this question, by phone, to Muhammad Mohiuddin, director of the Cardiac Xenotransplantation Program at the University of Maryland School of Medicine. Mohiuddin performed the first and second heart xenotransplants on living human patients. "Nobody has tried," he said. "So it's very hard to say. By now, so much is known about pig physiology and the pig genome that it makes no sense to use anything else."

Though there are those who would prefer it to be anything else. Mohiuddin, a Muslim, has been approached by both Muslim and Jewish thought leaders. "A lot of these organizations have called me. About why I picked the pig. Of all animals." In Islam and Judaism, pigs are considered unclean. "They're *still* asking." To which he replies: "If you can fund me with two million dollars, I can try something else."

Would recipients of pig organs risk excommunication or some other form of religious censure? No, say religious leaders from all major faiths, ten of whom, in 2019, participated in focus groups at the University of Alabama at Birmingham. The consensus was that saving a human life would take precedence over any dietary restriction. "That's the argument I provide to these people," Mohiuddin said. "I'm not eating it, okay? I'm trying to save someone's life."

Technically, if not ecclesiastically, gal-knockout pork is edible. GalSafe pigs created by Revivicor—the subsidiary of United Therapeutics that ferried the pig hearts to Maryland for Mohiuddin's xenotransplants—have been FDA-approved for consumption. Meat lovers with an alpha-gal allergy may now indulge without worry. I tried to get a price check from Revivicor but was told the pork is not yet for sale.

The Alabama focus groups included two made up of patients (and parents of patients) who had heart or kidney disease that might lead to their needing a transplant. I was a little surprised to see no questions touching on the concern, however irrational, that a pig organ—a heart, in particular—would somehow imbue the patient with some essence of their porcine donor. Mohiuddin told me that was a concern of the first heart xenotransplant patient, David Bennett. Though seemingly not a serious one. "He said, 'Will I be going *oink oink oink* after this?' We can tell you," Mohiuddin added, "Mr. Bennett behaved exactly as he behaved before."

And so it is pigs, and pigs it will likely stay. "Now that everyone is working with pigs, it's difficult to deviate from that. A lot of investment has gone into genetically modifying the pig and keeping it clean."

Not just clean but "superclean," a technical term. The

superclean pig must be housed in a superclean pigsty, called a designated pathogen-free (DPF) facility. ClonOrgan's five hundred pigs live in a DPF facility, where they are regularly tested for forty kinds of viruses, bacteria, and fungi. This is in keeping with human organ transplantation, wherein donors are tested and must give a detailed medical and social history, lest they end up donating Zika virus or syphilis or hepatitis C along with their organs. You can't interview a pig, so instead you raise them in an isolated, fiercely controlled, antiseptic environment. That's what I've come to see.

By 11:00 a.m., we're on the outskirts of Neijiang. Around us now are low hills, white-tiled houses, and groves of orange trees. Oranges are to Neijiang as pandas are to Chengdu: a source of civic pride and unrelenting municipal decor. Here are statues of oranges and fiberglass roadside oranges that I mistook for bollards. Oranges can be seen in the metal filigree design of the lampposts. The citrus grown here is special, Yi says. "Inside is blood." I've got gene editing on the brain. I picture human blood spilling from a juicer, then realize that of course she means they're blood oranges, the variety with dark red pulp.

The orange groves are part of the reason this site was chosen for the DPF facility. Orange groves require little maintenance, meaning that fewer people—and their germs—are around. And absolutely no other pigs. By government decree, there can be no pigs but ClonOrgan pigs within six kilometers in any direction. Yi explains that because the Chinese government controls the land in China, it is easier to establish such rules and enforce them.

The driver pulls off to the shoulder. I see no building or parking area. Tall grass grows wild on either side of the road. Have we had a flat tire? "We're here," Yi announces. They lead me to a path made of cement pavers the size of pizza boxes. Weeds grow in the spaces between. A small red sign in Chinese tells us we are entering a biosafety control area. Thirty pizza boxes in, the pavers, the whole path, abruptly stops. We stop too. We stand for a moment in the grass and the blazing sun. Yi directs my gaze to an industrial-style building across a valley, maybe a quarter-mile distant, the ClonOrgan logo barely readable.

I'm so confused. "So we're going to bushwhack down this hill, through those trees and then up . . . ?"

Yi laughs. "There's a river down there." The river, I learn, serves as a natural moat on one side of the ClonOrgan pig facility. Access is additionally blocked by a stone wall that the company commissioned. "The Great Wall of ClonOrgan!" Pan laughs. He is trim and with mildly graying hair and glasses that ride high on the bridge of his nose. Enormous red Chinese characters are painted along the wall and translate to, more or less, "ClonOrgan Medical Use of Pig Farm." To our left and down the hill, a footbridge spans the river, ending at a high metal security gate. Video cameras are mounted on poles, and loudspeakers will loudspeak at you should you venture too near.

Yi points out a building slightly uphill from the pig facility. This is housing for the staff, seven in all. Most have biology or veterinary backgrounds, some are management. None are hog farmers. The workers remain on-site day and night for three months, whereupon a new group rotates in.

It is dawning on me that I have come a long way to not see pigs. Yi verifies that we are not going inside. However, the ClonOrgan "control center" is nearby, so I can see the pigs and the interior of the facility live, via video-camera feed. We head back to the car.

Security at the control center, or its lobby, at least, is nonexistent. There's no front door, no front wall at all. Four jolly pigs of the same size and material as garden gnomes welcome visitors. The floor gleams, and the tinkling, mild piano music of upscale Chinese lobbies plays in the background. The inside walls display professional-looking graphics that explain ClonOrgan's progress in xenotransplantation. A video monitor plays a loop of Dengke Pan being interviewed on Chinese television. It's a walk-through press kit.

Pan explains the DPF measures inside the facility. Every three days, everything inside is disinfected. The rooms use the same kind of airflow system that keeps bacteria in check in state-of-the-art operating rooms. HEPA (high-efficiency particulate air) filters, negative-pressure airflow. Pig feed is disinfected twice. It's irradiated before it goes in and then, once the portions are readied, it is exposed to ultraviolet radiation.

We cross over to a full-wall display of live feed from video cameras all around the DPF. There are the pigs, with their black tushes and heads, four or five to a roomy pen. Pigs being pigs. Nosing around, sprawling on the floor asleep. The pens are a curious contrast to the narrative of antiseptic rigor and the glossy sheen of the control-center lobby. The pigsty walls are scuffed. The slatted floor gleams, but only in the spots where a pig has urinated. In two pens, a scatter of turds is visible. I guess this shouldn't surprise me. You can't knock in a gene for using toilets. You can't hose out the pen every time a pig relaxes a sphincter. This is as superclean as you get with five hundred penned pigs.

On the way out, I notice that one of the happy fiberglass pigs has a small red heart affixed, like a brooch, to its shirt. Five truncated yellow tubes protrude from it. Coronary arteries! He's wearing his own resected heart.

On a hill near the ClonOrgan path to nowhere, there is a small octagonal wooden structure, open-walled like a gazebo. Before leaving to go back to Chengdu, we stop to visit it. A carved sign in Chinese is affixed to the eaves—a symbol of the company, Yi says. But it's not the ClonOrgan logo. I inquire about the meaning of the Chinese character. Some discussion follows. "Organ temple," Pan says, and Yi repeats it.

"And what goes on here? Is there some kind of ceremony?" I would like to attend. Pan consults with Yi. "People from town can come and sit."

Yi nods. "Old people sit here."

Whatever it is, it's a lovely spot with shade and a breeze. We sit for a moment and talk about pigs and people and transplantation.

"There's another way," Yi says at one point. "Which uses the pig to grow human organs. It's called chimerism."

A chimera is a mixture of two animals, a creature previously confined to mythology. I've not heard about this before: pigs that are literally part human. Yi's friend Liangxue Lai was part of a team at the Guangzhou Institute of Biomedicine and Health that, in 2023, used CRISPR to create a pig that grew a mostly human mesonephros. The mesonephros is part of an embryonic kidney, and this one consisted of 60 to 70 percent human cells. Pan volunteers that he, too, is doing work in this area.

They explain how it works. Start with a pig blastocyst—the earliest stage of an embryo, a few days and a few cells old. Using CRISPR, knock out some genes, such that the pig is now unable to grow, say, a kidney. Now there's an open developmental niche. If you introduce some human pluripotent stem cells—cells with the potential to develop into whatever they're instructed to become—this empty niche is where those cells will thrive. Because the competition has been eliminated.

Ideally, the cells would come from the person who needs the organ. Not only would the pig be growing a human organ, it would be growing that specific individual's organ, thereby precluding rejection. And the pig's immune system wouldn't reject the human organ growing inside it, because the organ has always been a part of that pig. People who could afford it could pay to have a pig—or a series of pigs over their lifetime—growing organs genetically identical to their own, ready to go if and when they're needed. It's the "personal pig" Jeremy Goverman mentioned in the previous chapter.

And pigs could function normally with human organs? Hopefully so, Pan says. "But we are still in the laboratory stage."

"In 2010," Yi says, brushing away a fly, "a mouse grew a functioning rat pancreas."

When I got home, I emailed Liangxue Lai. I asked him when, in his estimation, chimeric organs would become a clinical reality. He said he could not possibly make a prediction, as it was as yet not even known whether primitive structures like the human mesonephros he grew could and would develop, in a pig, into a mature organ with normal function. One issue was the discrepancy between human and pig gestation periods—nine months versus four months.

The process is not as clean as one would hope. Human cells turn up not just in the targeted organ but in other organs as well, including the brain. Ethicists have raised questions. What if a chimera had enough human brain cells to enhance its intelligence and foster self-awareness? At what point would different moral standards apply? When would the animal have to be treated more like a human? I asked Lai about this. He replied that he had observed very few human cells in the pig's central nervous system. He added that researchers could, as a precaution, genetically engineer embryos that lack genes crucial to the formation of neural cells. But what kind of problems would that, in turn, create? I thought of my father-in-law, Bill, who in his last years took five prescription drugs. Two of the drugs treated medical conditions, two treated the side effects of those drugs, and one was for the side effects of a drug that treated the side effects of one of the original drugs. The messy daisy chain of medical progress.

We're in the car again. Yi sits in back with me. Yi, I say. Wouldn't it be so much easier to just convince people in

China—and more people in the U.S.—to donate their organs? I mean, this is China. The government could simply require it. Change the law, or just change the rhetoric. Make it people's patriotic duty to be organ donors. At the very least, take the halfway step and mandate an opt-out organ donation policy, as some European countries have done. Rather than asking people to opt in—by registering as organ donors—an opt-out system puts the burden of action on those who don't wish to donate. The default is for consent, not denial.

Yi doesn't think that's likely to happen, and not just for the sake of ancestors. Confucian tenets include reincarnation. She lays out the thinking. "You need to go to heaven with an intact body, because after maybe thirty years, you are born as a new baby." She laughs. "You don't want to be born as a new baby without kidney, without liver, without eyes!"

Also, after so many years of using the organs of death row prisoners for transplants, relinquishing an organ may be hard-set in the ethos as a punishment rather than a gift. And of course there remains the problem of rejection: the need for immunosuppressive drugs and the vulnerabilities and damage those drugs create. It's just too bad, because persuading humans to donate organs, rather than genetically altering them or growing them in pigs, seems so much simpler.

"Okay." Yi adjusts her shoulder belt and turns to face me. "Here is something simple." It is Yi's own work, and it involves xenotransplants of pancreatic islets, in order to treat type 1 diabetes. Again the parts come from pigs, but neither the parts nor the pigs are genetically engineered. The islets are "encapsulated."

First, a quick review of islets and type 1 diabetes. Islets are cell clusters within the pancreas that monitor blood sugar and secrete insulin to clear it when the levels get too high. (This is important, because unstable glucose levels cause damage to blood vessels, and over decades that damage compromises the circulation, leading to wounds that won't heal, amputations, blindness, heart problems.) In type 1 diabetes, one's immune system attacks one's own islets, creating a need to take over their work; diabetics rely on external or implanted devices to monitor their blood sugar and on synthetic insulin to clear it.

In the U.S., the company Vertex has been testing encapsulated islet cells sourced from embryonic stem cells. Encapsulating the cells would hide them from the deadly blunderings of the immune system. Yi's approach is to use encapsulation to hide pig islet cells from the immune system. Infinitely cheaper—it's pig offal!—and simpler.

Yi's background is in pharmaceuticals, so she knows how to build capsules. The islet capsules, each of which holds 50 to 100 of these cell clusters, consist of an extremely fine mesh. The mesh is made up of polymers cross-linked in a manner that creates tiny, precisely sized holes. Insulin can exit through the holes, and nutrients and oxygen can enter, but the cells of the patient's immune system are blocked. Access denied. Yi says it's working, that the diabetic monkeys in her lab required neither insulin injections nor immunosuppression for three months. Clinical trials were to begin soon.

If it turns out to work in humans, it will be a monumental achievement. Researchers have been working on encapsulation for thirty-plus years now. It may be less complicated than xenotransplantation or chimerism, but it isn't simple.

I checked back with Yi, via email, a year after I'd visited. She summed up the xenotransplants to date: two hearts, three kidneys, two livers. All but one patient had died within two months. Yi said that while hyperacute rejection had been prevented, it appeared that some sort of systemic, whole-body reaction was taking place. "I think, still a long way to go," she said. Yi compared the human body to a spiderweb. It seemed a lovely and apropos metaphor: something surprisingly robust and at the same time fragile, everything interconnected by strands we may not even see—and disturb at some peril.

Let's take pigs and capsules and chimeras out of the picture for a moment. What if it were possible to extend the shelf life of the ordinary human organs that become available for transplant every day? Could we one day bank hearts and kidneys the way we bank blood?

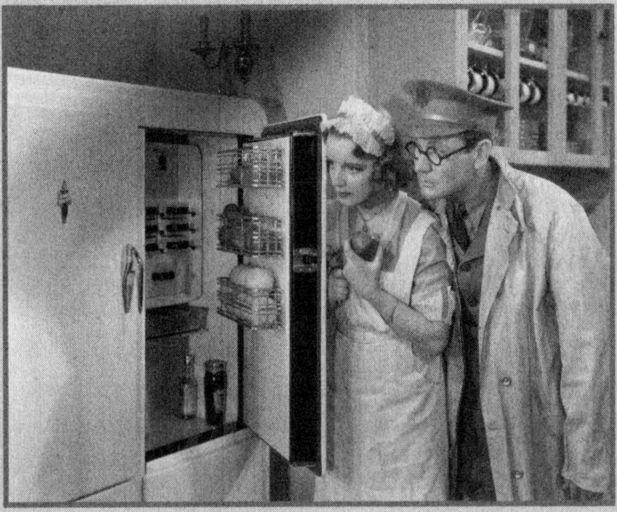

4
Heart in a Box

Creating Ultra-Long-Life Organs

Bob Bartlett is retired, but it's hard to tell. On any given day, you may find the eighty-five-year-old in his office at the Extracorporeal Life Support (ECLS) Lab, in one of the University of Michigan's three large medical research buildings. One of the many things learned at the lab under Bartlett's fifty-year watch is that it's possible to considerably prolong the lifespan of organs once they're outside donors' bodies. If, say, a heart could thrive on its own for days or weeks instead of hours, you could use that extra extracorporeal—or "out-of-body"—time for something other than rushing to get it transplanted. You could run some tests, so that the choice of whether or not to use the heart could be based on its function rather than on the donor's age. And if it weren't working well, you'd have time to fix it—to repair a valve or put in a stent—or even, looking ahead, tweak the genetics to make it a better match for the patient.

Bartlett is something of a metaphor for the lab's work and the inestimable value of extra time. Like those hearts, he may not be in prime condition, but it would have been a terrible loss to set him aside just because he was sixty-five. His formal status is "active emeritus"—retired yet still acting as director of the lab, even though there's now a co-director, the patient and accommodating Alvaro Rojas-Pena.

Bartlett moves slowly, with a cane, but his mind seems always to be hurrying. His sentences often trail off at the ends, as though his thoughts have sprinted on ahead, refusing to be held back by the plodding slowness of human speech. After a whistle-stop tour of the facility, Bartlett deposits me in the laboratory's operating room. One of the lab's fellows, a surgery resident named Wyeth Alexander, is removing a heart. Writing now, I mainly recall Wyeth's eyes, maybe because they're a nice dusky blue, or maybe just because his appearance at the moment was limited to what was on view between his mask and his scrub cap. Bartlett makes introductions. Wyeth doesn't offer his hand for a handshake, because it is underneath a beating heart. He's detaching it from its moorings inside the chest.

In almost every way, from the neatly arranged instrument tray to the vital-signs monitor overhead, this is a textbook surgical scene. The obvious difference is the inch and a half of pink snout peeking out from under the surgical drapes. Pigs are used for this work because their hearts approximate ours. I understand if you feel bad for them.* *Your heart goes out to them,* I almost typed, then stopped. Because really it's the other way around.

After it's removed, the heart will be hooked up to an experimental pump circuit gizmo in the next room. The focus of the study underway here is to see whether pumping blood through a donated heart at a lower flow rate might be less damaging to it. As a time-buying alternative to icing organs, hospitals can now buy portable perfusion units—so-called heart in a box systems

---

* By way of consolation, the University of Michigan, unlike large pig farms, follows protocols of the NIH Institutional Animal Care and Use Committee, the Association for Assessment and Accreditation of Laboratory Animal Care International, and the Institutional Animal Care and Use Committee for care, pain control, and humane dispatch. It goes without saying that saved lives justify the sacrifice more cleanly than do chops and sausages, which really don't at all. (I still succumb to carnitas tacos, but I would be thrilled if laws were passed to stop me.)

that keep a heart viable for eight to twelve hours, as opposed to four to six hours on ice. (The systems are expensive, used only in about one in twenty transplants.)

"They really blast it with blood," Wyeth says. Which would seem to make sense. More blood means more oxygen. But pushing blood through a heart under higher pressure hastens its decline. The capillaries start to leak and the tissues swell. Swollen muscle doesn't contract as effectively, and swollen valves don't close properly. Wyeth is looking for the sweet spot: a flow rate high enough to keep the heart sufficiently oxygenated but not so high that it causes damage.

Abruptly, everyone's eyes are on the vital-signs display. The heart is beating twice as fast as normal, almost more a quiver than a beat. Hearts have a built-in pacemaker called the sinoatrial node. Like the brain, it sends electrical impulses along nerve pathways to contract muscles. The heart's pacemaker causes its muscle cells to contract in a coordinated *lub-dub* manner. Wyeth uses the analogy of a mother keeping the kids in line. But the kids—the heart muscle cells—have their own electricity. "Every once in a while, there's a *Lord of the Flies*–type situation, where the kids begin to make their own rules and things go haywire." Arrhythmias, fibrillation. That's happening now. The heart's rhythm has been thrown off by the proddings and maneuverings of the surgery.

Fibrillation is acutely dangerous. There's not enough time between beats, between the *lub* and the *dub*, for the chamber to fill with blood. "So the heart is squeezing against nothing," Wyeth says without looking up. Blood pressure drops, circulation falters, death looms. "Arrythmias like this happen maybe ten percent of the time. Because you're here, of course it's happening today."

Wyeth flicks the heart with his finger. This is the open-heart surgery equivalent of smacking an old TV set to get the picture on track. This sometimes works to reset the rhythm, but not this time. He calls for the crash cart and calmly positions a set of spoon-shaped defibrillator paddles on either side of the heart. He looks like someone about to toss a salad.

"Clear." The body jolts, and the beat returns to normal. "Basically," Wyeth says, handing off the paddles to a technician, "we just slapped the cells across the face and stunned them long enough for the pacemaker to take back control."

Twenty minutes go by. Wyeth has the heart almost free of the body now. The next step will be to insert and secure the tubes that will connect the heart to the experimental pumping rig next door. Before this can be done, the heart has to be infused with something to stop it. A healthy disconnected heart will often keep beating on its own for ten or more minutes. After I got home, Wyeth sent me a video of a heart laid out on a blue surgical cloth, still beating twenty minutes after it came off the pump—and several minutes after Wyeth had sliced it in half like a deli sandwich roll. He has on occasion struggled with hearts that are still pumping while he's trying to shave off millimeter-thin samples of tissue to send to pathology. "It's very annoying for me," he says.

A perfusionist* named Joe Niman readies a teeny bottle of potassium, an electrolyte found in our cells. Among other duties, potassium regulates the heart's electrical signal. If you flood a heart with potassium, it disrupts and stops the beat. "If you ever want to kill somebody, this is more than enough to do it," Joe offers. "And it wouldn't leave a trace." When you die, your cells break down and their contents leak out. "So there's potassium everywhere anyway." Don't mess with Joe.

To hook up a heart to the team's perfusion rig requires a

---

* Perfusionists operate the machines that reroute and oxygenate blood during cardiopulmonary bypass surgeries. I had never heard of a perfusionist, nor had my phone, which dutifully changed it to "percussionist," surely the last professional you'd want in the OR working on your heart.

tedious session of cannulation: the marrying of tubes. Medical tubing is slipped inside five of the heart's severed vessels, the couplings then secured with sutures. A second surgeon, Dan Drake, has shown up to help. Drake is soft-spoken, with a suave, slightly formal manner and a thin white mustache. Every time someone addresses him—"Dr. Drake . . ."—my brain adds "Ramoray." I keep it to myself. He's probably not a *Friends* guy.

The chilled heart sits in a bowl while the two surgeons work the tubes. With no blood pumping through, a heart is a pale, floppy thing, a far cry from the familiar red graphic of emojis and Valentine's cards. This one, at the moment, reminds me of a skinless, boneless chicken breast. Drake holds a slice of the aorta and passes me a pair of surgical scissors so I can get a sense of how thick and rubbery it is. I would defy you to distinguish (without tasting) a ring of aorta from a ring of calamari.

After a half hour, the cannulation is complete and the heart is brought next door and connected to the pump. Tubes run everywhere, several stretching across the room at Chinese jump rope height. I have twice come close to bringing the whole thing crashing to the floor.

And that is the least treacherous thing about medical tubing.

The real danger is clots. When blood contacts a surface other than the inside of a blood vessel, it starts to clot* within milliseconds. If a clot breaks free and lodges in the heart or brain

---

* One of the largest clots on record assumed the shape of the entire right bronchial tree of one lung. It measured 6 inches long and 5-1/2 inches wide. Dumbfoundingly, it was expelled intact, during what must have been—quoting Wyeth Alexander here—"the most satisfying cough of the twenty-first century."

or lungs, potentially devastating things like strokes and heart attacks can follow. Clots form on medical tubes and devices because blood has proteins that recognize as "other" any surface that's not the lining of the blood vessels. These proteins stick to that foreign surface and attract platelets—makers of clots. Platelets release chemicals to attract other platelets and to create fibrin, the gluey essence of a clot. For this reason, people with any kind of implanted device that comes in contact with the blood—an artificial heart valve, say, or a stent—must take an anticoagulant or antiplatelet to freeze the process. Even the brief tenure of an IV tube during and after surgery demands some kind of antiplatelet. You can't really talk about extracorporeal life support devices without talking about clots.

One thrust of Bartlett's lab has been to understand how it is that blood doesn't clot on the surfaces of the blood vessels—and to transfer that talent to medical tubing. Some twenty years ago, a former ECLS researcher discovered that the endothelial cells that line healthy blood vessels secrete nitric oxide, and that nitric oxide knocks platelets out of action—briefly and only there. Once they're back midstream, the platelets work normally again. Bartlett's group has figured out how to incorporate nitric oxide into tubes and catheters, buying patients up to a week without risk of blood clots. That means many patients, post-surgery, may not need to take an anticoagulant. That's good, because drugs that prevent clotting raise the risk of internal bleeding.

For centuries, clots had hampered progress on blood transfusions,* which were carried out directly from donor

---

* With the 1914 discovery of the anticoagulant sodium citrate and the pressing demands of world wars, public blood donation began to go mainstream. Whereupon the new challenge was human squeamishness. Up through the 1940s, donor centers would set up chairs alongside arm-sized holes cut into a wall-sized partition. Donors would slide their arm through a hole, making a donation without ever having to see the blood, the needle, or the phlebotomist sitting on the other side. Blood bank glory holes!

to patient by a simple tube. Owing both to clotting and incompatibility,* results were at best inconsistent and at worst fatal, leading some practitioners to experiment with transfusions of other body fluids. "In consequence of the great percentage of death in general practice from the effects of the transfusion of blood, a new operation has of late years been proposed," announced John H. Brinton in a lecture at Jefferson Medical School in 1878 "That of intravenous injections of milk." This was the year of the milk transfusion. "Getting to be quite the fashion," remarked John Bryson in the *St. Louis Courier of Medicine*. "I would be false to my own convictions if I did not predict for 'Intra-venous Lacteal Injection' a brilliant and useful future," crowed T. Gaillard Thomas in an 1878 issue of the *New York Medical Journal*.

Practitioners disagreed about whether cow's or goat's milk was most efficacious,† but all were in agreement that the milk must be scrupulously fresh and warm from the teat. Thus began a brief, surreal chapter of medical history wherein livestock were ushered inside the homes of moribund patients. "A goat was brought into the pantry," Dr. Brinton began, laying out a case study for his audience. "The cow was brought into the room

---

* Between blood types, I mean, not patients. Though in the 1600s, when Hippocrates' theory of the four blood humors still held sway, transfusions were occasionally given as a kind of self-improvement or even as couples therapy. Write Leo Zimmerman and Katharine Howell in a 1932 *Annals of Medical History*, the seventeenth-century physician Johann Elsholtz "proposed that the temperament of the melancholic be corrected by transfusing him with the blood of a sanguine person . . . and he even suggested that marital discord be settled by reciprocal transfusions of the husband and wife." Would that it were so simple.

† Judging from the Letters page of the September 1917 issue of the Baldwin Park, California, newspaper *The Goat World* ("Devoted to the Upbuilding of the Goat Industry"), goats themselves are enthusiastic fans of goat's milk: "We have a young Saanen doe who sucks herself."

adjoining that in which the patient lay," recalled Dr. William Pepper* in an 1878 issue of the *Medical Record*.

Hospital inpatients posed more of a challenge. Edward Hodder, writing in *Practitioner* five years earlier, describes using an "old shed on the hospital grounds" for a series of experimental milk transfusions for cholera patients. "Everything being ready, I ordered a cow to be driven up to the shed and while she was being milked into a bowl . . . I opened a vein." Though one patient "collapsed . . . and died during my absence from the shed," two rallied and recovered. Hodder and a colleague "then applied to the Corporation for a good cow." The request was refused and the pair resigned, to the Corporation's likely indifference.

Medicine came to its senses quickly. Brinton himself concluded his paper with the speculation that it may have been the added volume contributed by the milk, rather than any specific "lacteal" element, that underlay a patient's improvement—the extra fluids staving off hypovolemic shock. And that "the injection of simple salt and water" would serve, as saline does today, just as well.

It's time to reanimate the heart. I would have assumed that some variety of medical-electrical jump start is in order. Not so. "The heart wants to beat," Wyeth says. "You just give it what it needs." Blood.

Within a minute, the heart is chugging. Hanging from the rig by its aorta, beating for no one. Beating for science. It's

---

* Not the Dr. Pepper of soft drink renown. That is Dr. Charles Pepper. Who was in fact a doctor, but not the inventor of Dr Pepper. Dr Pepper was invented, in 1880, by a drugstore owner named Wade Morrison, who was in love with Dr. Charles Pepper's daughter. Our William Pepper, I learned from the biography *William Pepper, M.D., LL.D (1843–1898)*, was a doctor from Philadelphia. As was his father, William Pepper, "known in the annals of medicine as the elder Pepper."

deep red again. The eye is drawn to it, as to a fire in a fireplace. Dr. Drake offers to let me go "don gown and gloves" so I can feel it.

Cupping a warm, beating heart in your hand is as surreal as you might imagine. It's the stuff of slasher films and human sacrifice, but to me it doesn't feel disturbing. Is there a word for awe mixed with tenderness? That's how it feels. It also makes me aware of my own heart, which is beating slightly slower than the one in my hand. I wait for a minute to see whether they'll start to synchronize.

"Mary?" The gentle tones of Dr. Drake.

"Yes?"

"All done now?"

I withdraw my hand. "Sorry."

Wyeth adjusts a cannula. "So we're doing a twenty-four-hour heart run," he says, sounding like a participant in an especially grueling heart association fundraiser. Twenty-four hours is an impressively long lifespan for a heart outside of a body. Reducing the blood flow rate is only one intervention the team has been testing. They're also infusing hearts with plasma (or components thereof), rather than whole blood, and filtering out wastes and fluids. Down the line, they may experiment with combining a period of off-pump cold storage, followed by a stretch of time with warm pumped blood, during which the heart could be assessed and, if needed, repaired. Much later, they'll test how well any improvements carry over to a heart that's actually been transplanted into an animal.

When I return after lunch, an echocardiogram machine has been wheeled in. The images it displays allow the team to gauge a heart's efficiency—how much squeeze the muscle is capable of, how much blood it's pushing out. "We can see how thick the septum is—which tells us how badly it's swelling," Dr. Drake is saying. "We can look at the valves and see how well they're working." On the display, he points out the mitral valve, named for its resemblance to a miter, the fancy-dress headgear of bishops and popes.

One by one, the surgeons and nurses peel off their papery outerwear and leave the room. A surgical resident will stay here overnight, monitoring and taking data, just him and the beating heart. He concedes that it is "a little Poe-y."

I say goodbye to the heart and its minder, and walk back down the hall to find Bartlett. We sit in the conference room, eating potato chips left over from a lunch meeting and chatting about the future of organ banking. He envisions hearts available to all who need them, hearts that have been evaluated and repaired, like reconditioned iPhones, ready for transplant. Hearts gene-edited to be a close match for their recipients, so they can get by with a lighter dose of immunosuppressive drugs.

"At the same time, we're trying to figure out what's important to add to make an organ run forever."

This is the first I've heard of this.

"The so-called vitalin project," Bartlett adds.

He explains. Organs have basic needs: oxygen, electrolytes and nutrients, and some way of removing waste products. You would think that you could hook up a heart to a rig like the one I just saw, which can meet those basic needs, and it would keep going indefinitely. But it will not. For reasons that remain a mystery, a heart needs a brain. When a person is declared brain-dead, their heart—along with their other organs—begins to fail after around twelve hours, even if the body is being oxygenated on a ventilator. The blood vessels lose tone, the capillaries start to leak, and within forty-eight hours it's all over. "An unpreventable sequence of dysfunction and failure," Bartlett called it in a 2004 *Journal of the American College of Surgeons* paper entitled "Vitalin: The Rationale for a Hypothetical Hormone."

But if you were to hook up an isolated heart to someone else's circulatory system rather than to a perfusion machine, it would beat far longer—at least three days. We know this because it has been done, here at the ECLS lab, with those most mellow of barnyard creatures, sheep.* The brain, Bartlett believes, is secreting something that's necessary to keep the organs functioning. Though he has not yet isolated the something, he has named it. Vitalin, by the way, is pronounced VI-talin, with the emphasis on the first syllable. As a child of the Vitalis Hair Tonic years, I keep saying vi-TAL-in, which has probably been irksome for Bartlett.

Also probably irksome was my asking, mostly but not entirely in jest, whether vitalin is perhaps a special substance produced by sheep. Bartlett patiently explains how we know that the organs-need-a-brain situation holds with humans too. This is a little dark. If a head injury or illness were to leave a pregnant person brain-dead, on a ventilator, their organs would keep functioning for the duration of the pregnancy. But once the baby is delivered, the familiar cascade of organ failure would begin. The baby's brain appears to be producing something that sustains the organs of the mother.

The reverse scenario also holds. "Very rarely," Bartlett continues, "a fetus develops without a brain." The "acephalic" baby's organs will continue to function as long as it's inside its mother's womb. "As soon as that fetus is delivered, it dies." Bartlett says. "You need one brain or the other." He reasons that there's a necessary hormone, or possibly several, produced in the hypothalamus, inside the midbrain. "The hypothalamus consists of hundreds of clusters of hormone-producing cells. We know what four or five of them do, but not the rest."

---

* Sheep may be mellow, but they are herd animals and isolation stresses them. This explains the unusual decor of the lab down the hall with the wraparound murals of life-size grazing sheep.

If vitalin exists, if this elixir of organ longevity could be synthesized or collected and bottled,* the benefits could extend beyond the realm of transplantation. Organs and glands could be put to work churning out medically helpful secretions. Your family and friends could come visit you as a liver or a pancreas, sitting there in your perfusion box, sipping vitalin and cranking out elixirs for strangers.

Dr. B., as everyone here calls Bob Bartlett, is best known for pioneering ECMO—extracorporeal membrane oxygenation, a critical-care technique that relies on a machine to oxygenate and pump a patient's blood outside their body (and haul away carbon dioxide). ECMO can take over the work of the lungs and/or the heart when they're not keeping up on their own. ECMO hit the news during peak COVID times, but it has been around since the 1970s, mostly for premature babies whose lungs haven't finished developing and aren't yet fully functional. (The alternative, for preemies, is a positive-pressure ventilator, which blows air into the lungs and can easily damage their delicate tissues.)

Bartlett developed ECMO while working as a surgeon and overseeing the lab. It was a lot. His wife, Wanda, recalls the morning she woke to find the couple's ailing golden retriever lying dead by the bed. "He stepped over it and went to work," she said during a dinner with me and a few of the lab's researchers. She meant it as a funny anecdote, but the message was clear. Bartlett's heart beats for his work.

As ECMO machines become more portable and affordable,

---

* The trademark Vitalin is up for grabs in the United States; however, since 1953, Vitalin has been a pet food brand in the UK. Thus anyone who searches for Dr. Bartlett's discovery online will immediately encounter marketing materials for Vitalin Beef and Veg Jerky Bites and Vitalin Chicken w/ Thyme and Root Veg.

versions of them may start to be used outside hospitals, in emergency cardiac care. Demetris Yannopoulos, founder of the University of Minnesota Medical School's Center for Resuscitation Medicine, makes emergency house calls with an ECPR (extracorporeal cardiopulmonary resuscitation) truck that Bartlett calls the ECMObile. Patients can be hooked up within minutes via the vampiric opening of vessels in the neck. Time is everything. If the hookup is done more than four minutes after the person's heart has stopped, the brain, starved of oxygen during that time, will likely suffer permanent damage, possibly to the point of leaving the patient in a vegetative state. They'd still be alive, but you wouldn't really say their life had been saved.

ECMO machines may eventually show up in rehab centers or even homes, to help patients who need a heart or lung transplant but are too compromised to withstand major surgery. Hooking them up to ECMO gives the heart and/or the lungs—the whole patient, really—an opportunity to rest and recover a little. And since they're not on a ventilator, intubated, they can talk and swallow. They can eat normally. Unlike you and me, they can do it without ever taking a breath. If they felt like it, they could watch Netflix with their head underwater. With sufficient nursing support, ECMO patients can even walk on a treadmill. Bartlett called it "sports ECMO" but they're hardly running laps. In a 2016 study of the safety of ambulatory ECMO, patients covered an average of 200 feet per session. Happily, no one tripped and "decannulated." The hope is that patients will be able to regain enough strength to become eligible to receive an organ for transplant. And if not, ECMO can ease the exhausting struggle to take in enough oxygen, improving their quality of life in the time they have left.

I tell Bartlett I've been following the progress of a team of researchers working on a different variety of non-lung breathing. Enteral ventilation via anus (EVA) delivers oxygen through the mucosal layer of the gut, making use of the rectum as a sort of third lung.

"We tried that," Bartlett says, crumpling a chips bag. "Doesn't work." The intestinal tract, he adds, has only about 20 percent of the gas-exchanging potential of the lungs. Bartlett's team had tried it from the reverse direction. "You can put a tube into the stomach and just blow oxygen through. You get rid of the CO_2 that way, too. Have it come out the rectum."

"Basically exhaling through your anus, then."

Bartlett says nothing.

"So, like, constant copious farting."

Finally a smile. "Doesn't seem very attractive."

The EVA team, led by Takanori Takebe, a professor with Cincinnati Children's Hospital Medical Center, found better results using a liquid—oxygen-saturated perfluorocarbon*— that also clears carbon dioxide. As of late 2024, the technique is in clinical trials. If approved for use, the perfluorocarbon enema may find its way into neonatal intensive care units as an alternative to ventilators, which easily damage an infant's lungs. EMTs and combat medics could also use it as a way to buy time in situations where ventilators aren't available or the lungs are too damaged to work properly.

While the intestine can't entirely replace the lungs, it does work surprisingly well as a stand-in for other parts and pieces. As we'll see, it's quite a versatile organ.

---

* Oxygen is extremely soluble in liquid perfluorocarbon. In 1966, researchers Leland C. Clark Jr. and Frank Gollan reported that perfluorocarbon holds up to three times as much oxygen as blood or air—so much oxygen, in fact, that a rodent completely submerged and "breathing the liquid" will survive on average for four hours. The experiment, published in *Science*, was carried out with the hope that the technique might prove helpful in, among other things, submarine escape. It was not to be. Necropsies of the animals showed damage to the animals' lungs in the form of "red areas distributed in a polka-dot pattern." This is not to be confused with the white polka dots that appear on X-rays of vertebral hemangiomas, collectively known as "the polka-dot sign." The seminal paper on the topic, "The 'Polka-Dot' Sign," includes, for clarity, a photograph: "Fig. 2: A polka-dot dress."

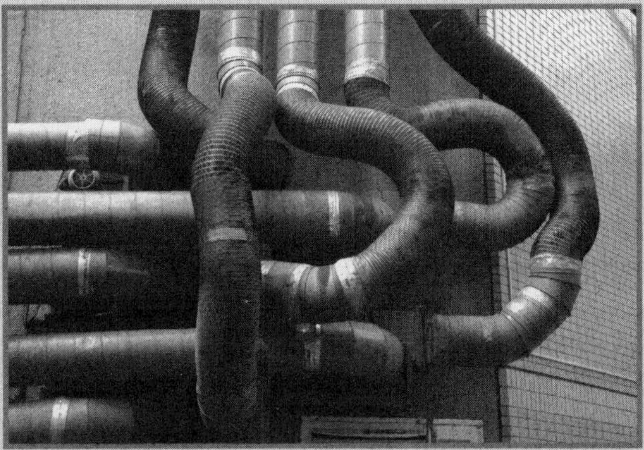

5

The Vagina Dialogue

Repurposing Your Parts

Locanda Veneta is an old-school Italian restaurant in the shadow of the Cedars-Sinai urology building, in Los Angeles. It is quiet, softly lit, and a bit of a splurge. It's the kind of place you might take your date for a romantic meal, especially if your date is, as mine is, a urologist. I've reserved table 12, a cozy corner two-top where most other patrons can't see or hear you, and the banquette is just long enough for two people to squeeze in side by side. It's the table for canoodlers, or people having an unrestrained conversation about surgically fashioning a vagina out of the colon. The banquette is good, because later my date, Maurice Garcia, will be coming to sit beside me with his iPad, so he can show me a video of a surgery carried out across the street at Cedars-Sinai Medical Center. Dinner and a movie.

Most of Garcia's patients who undergo this surgery are trans women; Garcia is the director of the Cedars-Sinai Transgender Surgery and Health Program. In particular, they are trans women who've had complications after a neovagina was created in the more standard manner—out of the patient's gutted and inverted penis. Garcia, with help from a colleague—a colorectal surgeon—pioneered the surgery in 2017. It interests me as an example of the remarkable and sometimes surreal adaptability—the *agreeableness*—of the human body. Here we have an organ that evolved for a specific purpose being drafted into service for a wholly unrelated one. The intestinal tract seems especially agreeable. Colons are recruited to serve as bladders and esophagi. The appendix has stood in as urethral tissue and helped rebuild a voice box.

I find equally remarkable the inventiveness and confidence of surgeons who dream up operations like this one. Who looks at the human digestive tract and thinks, *Moist, tubular, stretchy . . . Might that make a reasonable vagina?*

Garcia's team was the first to transform the ascending colon, but surgeons dating back as far as the 1890s have taken inspiration from, variously, the rectum, its upstairs neighbor the sigmoid colon, and the small intestine. "Dreyfus ingeniously made use of a hernial sac," marvels Francis Stewart in the 1913 *Annals of Surgery* paper "Formation of an Artificial Vagina by Intestinal Transplantation." Stewart's own patient, whose small intestine was recruited, wasn't a trans woman but rather a cancer patient—who, despite having been informed of the risks of surgery, stated "that she must have a vagina, or her husband would desert her."*

The sigmoid colon and the small intestine are still sometimes used today, but not by Garcia. The blood vessels that feed the sigmoid colon, he says, have a lot of variability and sometimes tether it to its original position too closely. The small intestine isn't wide enough. But the ascending colon is just right: nicely accommodating and fed by a longer and less variable artery.

I assumed Garcia's team did a "wet run" on a cadaver, but this wasn't the case. "I talked to my colorectal surgeon, and

---

* Stewart, unsure, had asked five of his colleagues to examine the patient and weigh in. "All unhesitatingly took sides with her," he writes, "one of the gentlemen stating that any operation destined to preserve the marital relations and keep the home intact was not only justifiable but mandatory." And here we see how the anus can be used as a substitute for the male brain.

he agreed: It makes sense, let's try it." A waiter arrives with menus, bulky as hardback books. Garcia thanks him and turns back to me. "Colorectal surgeons are used to cutting all sorts of pieces of bowel for all sorts of things—with all confidence they can put it back together." The pair consulted with the patient, explained the possible complications and side effects, and obtained consent. Unlike drugs and medical devices, no federal regulatory system exists for surgical procedures. For a radically new procedure, surgeons would typically apply for IRB (institutional review board) approval. But if the surgery, like this one, is a modification of an existing procedure, a team may proceed without that.

How do you balance the advantages of a workable vagina against the effects of losing eight inches of colon? The colon—aka the large intestine—is essentially a conveyor-belt food-dryer. The stomach delivers to the intestines a slurry of liquefied food, called chyme. From this slurry, the small intestine absorbs nutrients that the body can use and then passes the leftovers on to the colon, which pulls water from them. (I'm simplifying, of course.) Ideally the leftovers arrive at the rectum with a consistency that is wet and soft enough to easily expel but firm and dry enough to hold in until you're situated to let it go. In the words of the Bristol Stool Scale, "smooth and soft, like a sausage or snake." If you shorten the conveyor belt, less liquid gets absorbed. The more liquid the material, the harder it is to hold in.

Garcia says that a deficit of eight inches is not enough to create a scenario of chronic diarrhea. The complication he does see is not in the digestive tract but deep inside the neovagina. Some patients develop bleeding and burning—a kind of colitis of the vagina. The burning sensation was unexpected, because the colon in its birthplace doesn't perceive that type of pain. (The colon registers pain via receptors that sense when the organ is being stretched—by the pressure of intestinal gas, say, or the movements of a colonoscope.) Garcia's theory is that bowel tissue in its new location might be missing some necessary nutrients that it normally derives from "the fecal stream."* He suspects that what's lacking might be short-chain fatty acids. Milk products are a good source, but how do you get them up there?

"I always preface this by saying, 'I'm going to tell you something, and it's going to sound really weird.'" Our waiter is back, to recite tonight's specials. He waits for a pause in the conversation. What Garcia tells the patients is to douche with half-and-half once a month. The waiter maintains a neutral silence. "It seems to help."

Based on vaginoplasty papers I've been reading, the colon has an advantage over penis skin. From a paper in *BioMed Research International*: "The mucosa feels and appears more like vaginal mucosa with the added benefit of self-lubrication." (The intestine is sufficiently vagina-adjacent that the authors of a pamphlet on ostomy aftercare felt the need to warn patients that their stoma—the open hole through which the intestine empties—"should not be used at any time for penetrative intimate moments.")

Garcia doesn't agree with the lubrication bit. "That's written by the surgeon, not the patient." He explains that the mucus the colon produces isn't the same as the vaginal secretions of sexual arousal. It's constant. "If they don't douche frequently enough, it builds up. And it's thicker. More clumpy. It's . . ." Garcia searches for the right word.

"Grumous?" (When I first got on Twitter, I used to post a Medical Dictionary Word of the Day. *Grumous*, adj.: thick, clotted.)

"Mary, what sort of physician are you? What specialty?" Garcia must have assumed, when I first wrote to him, that I have an MD. I do not. This I tell him. There's a pause that is hard for me to read. Annoyed? Disappointed?

---

* A particularly lovely medical phrasing. You can almost imagine spreading a picnic blanket or casting a line.

Merely distracted. The waiter has set down bread and a small dish of pesto. Garcia takes a slice of focaccia and goes into it. Spreads it like a bricklayer. I follow.

He finishes, brushes crumbs from his fingers, and leans back. "Sorry, I hadn't eaten all day." Surgeons, as a breed, skip lunch. When your day's work lies unconscious with its hide opened up, there's never a good stopping place. "Where were we?"

"Self-lubrication. Not a selling point."

A short laugh. "Yeah, my patients are not like, 'This is so great, I don't have to buy lube!'"

I have a question about peristalsis: the squeezing wave of contractions that moves the contents of the intestines along their journey. "So this piece of colon, when you move it to its new location, is there still peristalsis?"

Garcia pauses for a sip of Chianti. "Yes, it definitely peristalses." That's another reason to use the ascending colon, he says. So the peristalsis happens in a helpful direction, with mucus moving from the deepest end of the neovagina outward to its opening.

I was more interested in the stimulative potential of a peristalsing vagina. "That could be kind of fabulous for a partner with a penis, no?"

No. "It's not that aggressive," Garcia says. "It's subtle."

My final question has to do with the neoclitoris. This is a clitoris that can be made using tissue from the nerve-dense underside of the glans penis. For my third book, *Bonk*, I reported on a Frenchwoman, Marie Bonaparte, who in the 1920s studied the link between orgasmicity and vaginal-clitoral distance. She'd found that *teleclitoridiennes*—her coinage, meaning "women of the distant clitoris"—were the least likely to reach orgasm from intercourse alone. I wonder (aloud) whether surgeons like Garcia have considered a closer-than-natural placement of the neoclitoris. Or even, as in the anatomy of the pig, just inside the vagina.

"It's a good thought," Garcia says, but he doubts it. "Most trans women want to look normal." And while it's true that "normal" encompasses two-plus centimeters of variation, never is it sitting right on the rim of the vagina. And a freshly sutured wound in that location would make postoperative dilation uncomfortable. Trans women who have the surgery are instructed to follow a regimen of vaginal dilation. In Garcia's practice, it's three times a day during the first three months.

Garcia holds a patent on a dilator designed specifically for women who have undergone intestinal vaginoplasty. Colonic vaginas are longer than those crafted from penis skin, and most dilators come up short. His is also unique in that it serves double duty as a douching wand. It's hollow and comes with a turkey-baster-style bulb attachment. "It occurred to me," he says, "that since you're already sticking a dilator in there, why not make the dilator the douche?"

The waiter sets our pastas on the table. "Parmigiano?"

"A little bit," says Garcia. "Thank you." He leans out of the way. "So at the end of dilation, you douche."

Garcia owns a medical device company. Inventing is a hobby and a passion. It's also a reason the transgender field appealed to him. Because gender-reaffirming surgery is a relatively new area, there are opportunities for innovation. New circumstances breed new procedures, and then better procedures. Garcia mentions the work of Harold Gillies, whose pioneering techniques on soldiers maimed in World War I laid much of the groundwork for modern plastic surgery. Besides that, he finds the work profoundly gratifying. "Patients are very appreciative. I just fell in love with it."

What Garcia says next surprises me. A high percentage of his patients, he tells me, don't have vaginal intercourse. Then why go through with the surgery? "I think a big motivator is simply getting rid of the male anatomy. Just on a day-to-day basis, looking down there and seeing something weird and out of place is what's most bothersome."

The prostate, however, remains. Might that add pleasure—a trans woman's partner's penis stimulating the prostate during sex? Unlikely, in part because the relevant nerves are severed as part of the operation. Garcia twirls his pasta. "The other thing about the prostate, Mary, is that if women"—trans women—"start hormones early in life, the prostate . . ." He swallows. "There's nothing. It's like a peanut."

Because of the relatively low interest in vaginal penetration, a third of Garcia's patients opt for vulvoplasty alone. Vulvoplasty surgically creates an anatomically realistic vulva—labia, urethral opening, external clitoris—but no vaginal canal. "They don't plan to have sex, or they're happy with anal sex, or they're only interested in women." Garcia is supportive of the choice and is careful to mention it as an option. He tells his patients that female genital anatomy has two parts: what you see and what you don't see. "The bit you don't see is going to cost you a lot." He means not money but effort and sometimes trouble. "You have to dilate it, douche it. And if there are complications, that's where they're likely to be." He wipes his mouth with his napkin. "So if you're not going to have sex with a penis, or with anybody, forget the damn thing."

Not all surgeons mention vulvoplasty to trans women who come in for a consultation. I wondered whether this might have to do with the higher remuneration of a more extensive transformation. Garcia believes it has more to do with surgeons' perceptions. Male surgeons tend to have a ciscentric perspective. "'Oh, you want to be a woman? You need this and you need that. Let me tell you what you need.'" A president of Garcia's surgical specialty group once confronted him at a conference demanding to know why he was creating vulvas without vaginal canals. "He said to me: 'It seems unfair.' I'm like, 'What are you talking about? The patients *choose* it.'" The waiter glides into view. "They say, 'I'm not going to use it, why the fuck do I want to dilate it?'"

"Are you enjoying your meal?"

"It's delicious. Thank you."

I'm reminded of how, for years, male surgeons often failed to mention to mastectomy patients that *not* surgically reconstructing their breast(s) was an option. That instead of multiple surgeries and a one-in-three complication rate, they could simply, as the choice is now known, "go flat." It took breast cancer activists and multiple Facebook support groups—with help from comedian Tig Notaro's defiant, shirt-off 2015 monologue—to push open the door for that conversation and that decision.

When cisgender male surgeons let their sexual worldview dictate their practice, it's not only women who endure the consequences. Garcia has had at least four trans men who came to his clinic for a penis reduction because the excessive girth[*] of their new phallus made it impossible to penetrate their partner. "Some of our patients relayed that their final dimensions were larger than they expected or that they believed their surgeons took personal pride in their large final neophallus dimensions," Garcia and three colleagues wrote in "How Big Is Too Big? The Girth of Bestselling Insertive Sex Toys to Guide Maximal Neophallus Dimensions," a 2017 study published in the *Journal of Sexual Medicine*.

---

[*] We're talking about girth, not length, because girth seems the more salient dimension in terms of female sexual pleasure. The distinction seems to have escaped the authors of the journal paper "Nose Size Indicates Maximum Penile Length," who sought scientific verification of the saying "Big nose, big hose" (their phrasing). They report that nose size was "highly related to stretched penile length," or SPL. SPL was standing in for erect penile length, which the researchers were unable to ascertain, because all 126 subjects were dead. (These were men who'd been autopsied at the Kyoto Prefectural University of Medicine, whose institutional review board, perplexingly, approved the project.) There was no correlation between nose size and penile girth.

"Certain surgeons kind of like thump their chest that they made such a big one," Garcia is saying. He takes a last swallow of Chianti. "It's so stupid." (In fairness to surgeons, trans men fairly commonly, Garcia says, request a neophallus larger than the average natal penis. But the extreme cases seem to have been the surgeons' doing.)

And how big around, then, is too big? Garcia's team hypothesized that the "upper limit of receptive girth would be reflected in the upper limit of girth among insertive sex toys." So they went online and looked into it. They focused on the category "realistic dildos," those modeled from actual erect penises—as opposed to, say, novelty bachelorette-party dongs. Sex toy vendors, like any other online vendor, include product dimensions. "Like shirts," Garcia says, scanning a dessert menu. "They tell you the girth, the insertable length." Almost like shirts.

Too big apparently equals: anything much beyond 13 centimeters (about 5 inches) around. "Beyond that, there wasn't much available."

And how big would be just right? What should surgeons aim for? Garcia and his team answered this too. Using data from a company that tracks online sales, they looked at the dimensions of the top-selling realistic dildos. They also went into a Good Vibrations store and hand-measured fourteen realistic offerings in the "bestsellers" section. This is not an unusual component of in-person dildo shopping. "They're very accommodating," Garcia says. "You can go in and they'll give you a paper ruler."

The mean circumference for the bestselling realistic dildos sold online and in Good Vibrations is 5 inches. If a man goes a lot bigger, it shouldn't be the surgeon who decides that. Possibly it shouldn't be the man either—or not the man alone. "We always," wrote Garcia and his team, "encourage the involvement of long-term sexual partners."

Most of the surgeons Garcia has been referencing were plastic surgeons, not urologists. "I've seen penises made with a fibula. No urologist would ever do that." The fibula is the less substantial bone of the lower leg. Because people can walk without a fibula, it is sometimes used to rebuild the jaw and other bones in cancer patients. As an erectile aid, however, bone is a risky choice, as it can poke through the skin at the end of the neopenis during intercourse. (A neopenis lacks the fibrous sheaths that encase the twin erectile chambers of a natal penis.)

Garcia finds it appalling. "I can't help but feel there's a bit of an element of *Oh, they're just transgender patients, we can try anything.*"

In 2002, the Russian surgical journal *Khirurgiia* published a case report that would seem to support this. Translated, the title reads: "Finger Transplant in the Creation and Reconstruction of the Penis."

The author, a Georgian plastic surgeon named Iva Kuzanov, used a patient's own middle finger to reconstruct the man's penis, which had been lost to cancer. I could find only an abstract. A video clip online showed the initial cut into the skin of the finger, then stopped—well short of answering any of my questions. Was the skin left on? Were the tendons hooked up? Could he crook the penis like a finger?

Emails in English, Russian, and Georgian went unanswered. I decided to just go there and show up. I called my longtime travel pal, Steph. I hear the Caucasus Mountains are glorious in autumn, I said, and a surgeon in Tbilisi transplanted a finger to serve as a penis.

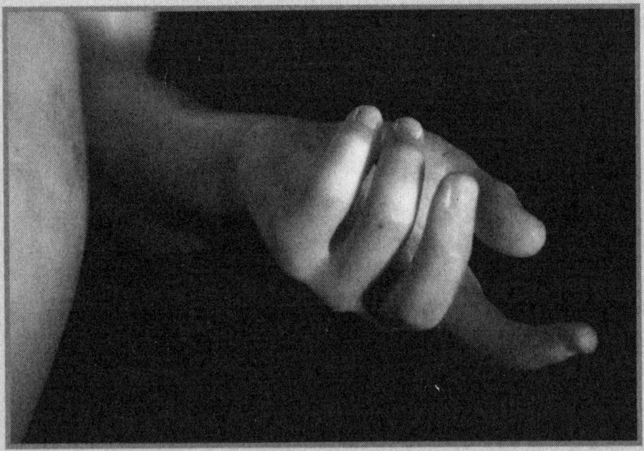

6
Giving the Finger

Some Transplants Are Tougher Than Others

The Georgian alphabet is a roly-poly, tassled affair, so visually pleasing that you find yourself taking photos of random strings of words. ("Under video surveillance," said a desk clerk when asked to translate a line of exquisite hand-lettering on the hotel exterior.) Right now, it's less enchanting, because I'm lost. A taxi has dropped me and Steph on a hectic Tbilisi street with few English signs and no street numbers in view. The Kuzanov Clinic is somewhere around here, but where?

As the taxi pulls away, Steph elbows me. She points to a billboard across the street and two stories up. It's installed perpendicular to the building facade, such that it extends out across a lane of traffic, the kind of placement normally reserved for freeway signs and other crucial information for drivers. The words are in Georgian, but I recognize the photo and the man from his website. White lab coat, dark glasses. Arms crossed, big watch. The elusive, not-so-elusive Iva Kuzanov.

In the building lobby, we ask the security guard which floor for Kuzanov. He holds up five fingers.

"Well, we know *he* hasn't had it done," Steph says as we step into the elevator.

A long white hallway leads to the clinic reception area. The lighting is fierce, and the floors are tiled in high-reverberation marble. There is no sneaking up on Iva Kuzanov. I introduce myself to a woman at the reception desk and hand her printouts of the letter I'd emailed, one in English and one in Georgian, prettier but likely Google-Translated into imbecility. She reads it, then ushers us to the waiting area and sits down with us. Her name is Nana, but she is no one's grandma. She's maybe thirty, with swooping eyeliner and perfectly executed brows. Her brown hair is pulled into a sleek ponytail. I am the usual frowsy mess that is Mary living out of a suitcase.

Nana asks if we've arranged a meeting, and as you know, we have not. Though not for want of trying.

"Dr. Kuzanov is on vacation." Not to return for two more weeks.

"But we've traveled all the way from America to see him." To which a fair and logical response would be, "Well *you're* a couple of idiots."

As I'd hoped, Nana takes pity on me. She invites us to wait in an office, presumably belonging to a different surgeon who attaches fingers to groins. On the desk are piles of folders and papers and an unusually large glue stick. "That's what they use," whispers Steph.

Nana returns with the surgeon. The conversation reveals that although this man has observed the procedure in question, only Kuzanov has performed it. Four or five times over the past fifteen years. "We take this finger," he says, touching the middle digit of his left hand, "with the phalanx and metacarpal bones." He traces a path from the tip of the finger to his wrist. "From here to here." There are photographs, but he has none.

"We can see on Dr. Kuzanov's computer," offers Nana. We follow the *pockety-pock* of her stilettos down the hall and up a flight of stairs.

Nana takes a seat at Kuzanov's desk and starts opening files on his computer desktop. After a few false starts, she locates a folder of photographs documenting a finger transplant performed on a cancer patient. Most of the man's penis had been amputated, leaving him unable to do two of the things most men like to do with their penis: penetrate and pee while standing. The surgery would restore both.

Before I arrived, I'd pictured a whole, unadulterated finger—skin, knuckles, fingernail, all of it—in the place where a penis would normally be. Though of course I knew that was unlikely. In reality, the transformation bears similarities to a standard phalloplasty: a flap of the patient's skin is removed from elsewhere and wrapped around a rigid interior implant. Here, instead of commercial penile implants supplying the rigidity, it's the patient's middle finger. The digit is skinned but otherwise implanted whole.

Nana clicks again. "This is the finger inside with the skin wrapped around." This photo shows a sort of skin log lying in a small surgical basin. I'm reminded of a Vienna sausage, a canned meat product I enjoyed as a child—same approximate color and circumference, same blunt ends. A second operation is done to create the glans and fine-tune the appearance.

"This is the result." The after photo. We lean in. Pretty sure that's stubble we're seeing.

Nana confirms this. "Yes, hair he has. But he can make laser. Like ladies." With phalloplasty, the skin flap is often taken from the underside of a patient's forearm, because that area is usually—but clearly not always—hairless.

"Here, an X-ray." This is a first: a human penis bone. Rodents, dogs, and pinnipeds like seals have penis bones—baculums—as do some primates, but not humans. I once profiled a researcher who studied baculums, and at the time we spoke, no one was certain what purpose they served. They seemed mostly of help to taxonomists, who use them to differentiate species that otherwise look the same. If, on a dig in some distant future, this man's skeleton is somehow unearthed, scientists may conclude that a unique species of hominid once roamed the Caucasus.

As with more standard phalloplasty, the urethra is threaded

through the neopenis. Now the man can pee standing up. The next photograph confirms this: the man, from the chest down, inside a shower stall, urinating.

Overall, it's an impressive result. If not for the stubble, I wouldn't suspect this was anything other than a garden-variety penis. Nana clicks on to the next image.

Now I would. It's a side view of the patient, the penis out straight until a couple inches along, where it makes a sharp upward bend. Like this: Lie your hand flat, palm up. Now bend your middle finger. Did he use his hand to position it that way, or can he actually crook the internal finger? I picture the man reclining on his bed, beckoning to his partner. But for that to be possible, tendons would need to be hooked up to muscles in the forearm. More likely, it's a helpful feature that prevents the man from having a permanently protrusive erection.

"Yes," confirms Nana. "So he can put on his trousers."

Before the development of inflatable penile implants, there were malleable—bendable—ones, akin to Gumby limbs. You would bend them how you wanted them, and they'd hold the position. But a man would bend his malleable unit at the base of the shaft, not halfway along. The idea was to push the whole organ flat, like a Murphy bed.

Nana clicks again. It's another photograph of the penis, now with a rustic ceramic pitcher hanging off of it. Exactly as you or I could put a hand out, palm up, and hang a pitcher off our middle finger. The pitcher is hand-painted, white with a red and green floral pattern. I'm picturing the man and his partner again, this time having lunch in the kitchen. *A little iced tea, my love?*

"It's not full," says the surgeon. He means the pitcher. "This is just to show the resistance. It's very strong." Effectively penetrative, I think he means.

So the man can have intercourse, but what does it feel like for him? In female-to-male phalloplasty, part of the clitoris can be sutured to the underside of the glans of the neopenis. But this man, the cancer patient, of course had no clitoris, and his nerve-rich glans and most of the rest of the shaft were discarded after the penectomy. Is it possible to achieve orgasm by stimulating the skin of the underside of your forearm?

The question triggers discussion between Nana and the surgeon. Nana turns back to us. "Permanently," she says. I don't know what she means. Later, I recall a paper I'd come across while researching an earlier book. Soviet surgeon A. P. Frumkin reconstructed the genitals of four Red Army soldiers using skin from their abdomen (wrapped around a piece of rib cartilage, not a finger). "At first," he wrote, describing the cases, "the absence of the glans penis results in very weak orgasms and ejaculations. In time, however, new 'erotic zones' apparently develop, and the orgasm approaches normal intensity."

Nana adds that the man was sixty at the time of the operation, and his wife was thirty. So possibly he had it done for her sexual satisfaction as much as for his own.

"She was very happy," Nana says.

Among the many "after" photos was a single image of the patient's hand, with its missing middle finger. Because the knuckle and adjoining hand bone were also removed, Kuzanov was able to close the gap, and the spaces between the three remaining fingers are even. One might not, at a glance, pick up on the irregularity. Still, it's a bit of a price to pay. I wondered: if you're committed to using a finger in this way, might it be better to transplant one from a deceased donor?

Of course, as we learned in chapter 2, that is a different, more problematic, class of transplant—not an autograft (tissue from your own body) but an allograft (from another person's body). The advantage of using one's own bits, of course, is that the immune system doesn't respond to the tissue as foreign and proceed to reject it. Many of those in need of a transplant are cancer patients, and suppressing their immune system would leave them more vulnerable to a recurrence. A finger is what's known as a composite tissue allotransplantation, meaning it comprises multiple types of tissue—skin, muscle, nerve—and thus the immune system's reaction would likely be more intense. Until around 2000, surgeons hesitated to undertake these kinds of transplants—especially since hand and face and penis transplants are not considered "life-saving," as are transplants of a heart, say, or a kidney.

One thing that encouraged surgeons to move ahead was a technique called marrow infusion. Transplant recipients can be given a dose of the donor's marrow—marrow being a source of immune cells—as a way of quieting their own body's immune response. The hope was that lower doses of immune-suppressing drugs could be used, and that that, in turn, would lower the risk and severity of immunosuppression's side effects: the infections and cancers that a tamped-down immune system misses, as well as the toxicity of the drugs themselves, which damage the kidneys over time. The discovery generated enough optimism that hospitals were starting to support "life-enhancing" composite transplantations.

Everywhere you turned, it seemed, you saw a news story about another medical first. In 2008, a double complete arm transplantation. In 2009, the first face, jaw, and tongue transplant. In 2011, the first double leg transplant.* While working on *Grunt,* around 2015, I watched a surgeon at the Johns Hopkins School of Medicine transplant a penis from one cadaver to another, as part of the research for the institution's first penis transplantation.

---

* If you buy into Christian miracles, the first leg transplant dates to the third century AD, when Saints Cosmas and Damian cut off the leg of a freshly deceased African man and transplanted it onto an ailing deacon. If the circa 1370 altarpiece *The Miracle of the Black Leg* is factual, the pair then replaced the dead man's leg, inside his coffin, with the deacon's swollen, gangrenous amputated limb. Thoughtful! The two saints were later decapitated, though not for transplantation.

And then the headlines began to peter out. Does the media only cover the firsts? Have these transplants become routine? Does Blue Shield cover face and penis transplants now? Or are they no longer being done? The surgeon I met at Johns Hopkins, Rick Redett, put me in touch with a colleague, also a plastic surgeon, at another research hospital. Branko Bojovic is a clinician investigator with Massachusetts General Research Institute, in Boston. He has years of experience with all the composite transplants: faces, penises, testicles, hands, arms.

"There's been some cooling off," Bojovic said when we spoke by phone. Even with the lower doses of immune-suppressing drugs, the side effects have persisted. He reeled them off: lymphomas, skin cancers, rejection episodes. Infections that the rest of us don't get. A sort of chronic vasculitis that changes the transplant and makes it less supple, less usable over time. Kidney damage remains a serious issue. Bojovic has a face transplant patient who now needs a kidney transplant. People with transplanted hands are asking for them to be removed. Face transplant patients don't have that option; some are waiting for re-transplants—a third face.

"It's been somewhat sobering in terms of what happens over time," Bojovic said. "We've done some incredible stuff, incredible technical feats, but at the end of the day it's been humbling—our inability to win the immunosuppressive battle. And to see how the human body seems on some level to always sort of win no matter what you do."

Insurance companies still don't cover these operations. The Department of Defense and hospitals themselves used to provide funding, but this has dwindled. "There's much less appetite," Bojovic said, "to allocate the resources for these procedures that cost two to three hundred thousand dollars."

At the same time all of this has been unfolding, impressive progress was being made in the field of prosthetic limbs. I had asked Bojovic about leg transplants, why no one other than one surgeon in Spain had ever attempted them. In part it's because it takes so long for nerves to regrow. By the time the nerves reach the end of the limb, the muscles have catastrophically atrophied. But it's also because prosthetic limbs are better than they used to be. So in order to justify the risks of transplantation, Bojovic said, the transplanted body part would need to "look, feel, and work much better than any state-of-the-art reconstructive method," including prostheses. With legs anyway, that's not the case. Why take on the risks of a lifetime of immune-suppressing drugs when you can replace your limb with a "bionic" prosthesis?

I learned recently that lens replacement surgery for cataracts has become so reliably safe, and the visual outcomes so precisely predictable, that people without cataracts—people who are simply, say, nearsighted—are having their lenses replaced. It caused me to wonder: Have prosthetic limbs progressed so far that people are opting to have a healthy but underperforming foot cut off?

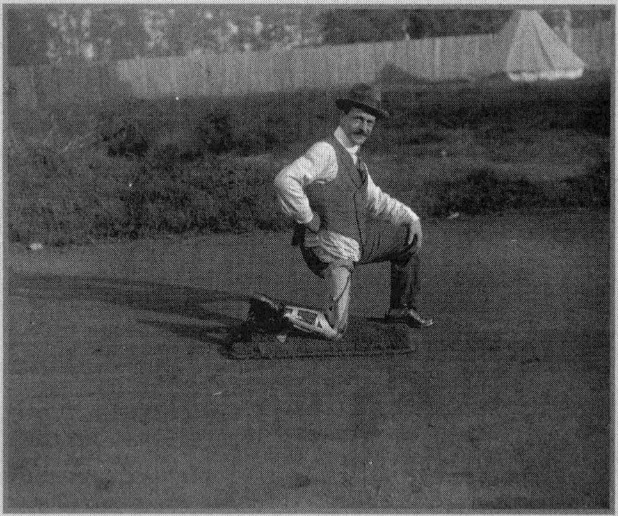

7

The Cut-Off Point

Longing for a Prosthetic Leg

The holdings in the guest library of the Marriott Desert Springs provide a sense of who typically stays here. *450 Best Sales Letters. Margaret Thatcher: The Downing Street Years.* The second edition of *Digital Customer Service.* This week is unusual in that no corporate meetings are going on. The golf course is empty, as is the business center, and seven of the nine guests in the lobby right now are missing a leg.

Six hundred rooms are occupied by attendees of the Amputee Coalition National Conference. I'm here because of one of my readers, Judy Berna. Some years back, Judy emailed me to suggest I write a book about pro football referees. In the back-and-forth that followed, Judy mentioned that she is an amputee. Specifically, she is an elective amputee. She asked to have her left foot cut off. Judy was born with spina bifida. After the removal of a tumor on her spinal cord, scar tissue entangled with the cord, and her foot twisted as it grew. Her childhood was dense with surgeries and stubborn infections. "My foot had become a science experiment," she wrote in her memoir, *Just One Foot,* "a mess of flesh and bones but few working nerves." It exhausted and limited her. Over the years, she'd watch people with prosthetic limbs effortlessly managing things she struggled with every day.

"Your hip joint wears out, you replace it," she wrote in an email. "How is this so different?" Yet it took decades to find a surgeon willing to amputate. When it became clear that I found all of this more interesting than things that happen on a pro football field, she invited me to join her at the conference.

In the registration line, we've been talking to Clayton Frech, who runs Angel City Sports, in Los Angeles. Frech and his company help amputees get into, or back into, athletics. When I bring up elective amputation, he mentions a client, an ex-marine whose injury had left him with foot drop. When you bring your foot forward while walking, you lift the front of it without thinking. With foot drop, nerve damage interferes, and as a result the toes drag. The man knew that a prosthetic limb could restore his gait and pace, but no surgeon would agree to cut off his foot, so he shot it.* "By accident." Frech applies air quotes. Now no one could say it was salvageable. The foot came off. "His life is exponentially better," Frech says, looping his name badge over his head.

"I had four kids," Judy says. "I couldn't shoot myself in the leg." Judy was raised in a religious family, which surely added to the pressure to accept her lot. Her father has the kind of deep-set

---

*   Outside of a few cases of psychosis, self-amputation is almost always undertaken in the face of a worse alternative. You may be familiar with the practice among the *yakuza,* the Japanese mafia, of *yubitsume,* or "finger-shortening." Chopping off part of one's own pinkie was done to show remorse for some kind of transgression, but on a more practical level, it was undertaken to avoid a worse fate—for example, execution or *seppuku* (self-disembowelment). I learned all this in an article by three neurosurgeons, in the always sunny *Journal of Injury and Violence Research.* It was interesting to have the surgeon's perspective. The authors noted that although the severed bit—the "proximal digit at the distal interphalangeal joint"—was traditionally presented to the offender's boss, some *yakuza* families would allow the man to take it back and hurry to the hospital to have it reattached. Others allowed anesthetic to be administered for the severing. *Yubitsume* has largely been replaced by a perhaps more dread personal diminishment: a cash fine.

faith that has him thanking God when the traffic lights go his way. Of Judy's four children, one has epilepsy, one has severe scoliosis and type 1 diabetes, and one has a metabolic disorder. All of that, on top of her own health struggles, worked to erode her faith. ("I guess God was busy with your traffic lights, Dad," she teased at one point.) Judy is tall and fit. She speaks in an upbeat alto, not full-on cheery but always with a laugh in its back pocket. She has straight wheat-blonde hair that she wears down, with bangs. Writing this now, I can't recall her outfit that day, but I do recall what her residual limb* was wearing, because the prosthesis was wallpapered with Seattle Seahawks logos. (Judy also owns a leg for autographs. Seahawks quarterback Russell Wilson acquiesced after the crowd fell into a chant: "SIGN. THE. LEG. . . . SIGN. THE. LEG. . . . SIGN. THE. LEG.")

Clayton Frech's field is known as adaptive sports. Often a sports prosthesis or "terminal device" is involved—"terminal" because it attaches to the end of a prosthetic limb. (Most people, and catalogues, use "TD," perhaps because "terminal device" sounds kind of awful and ominous, a name for a doomsday detonator or a euthanasia machine in some dystopian future.) I once went bowling with the Bay Area amputee group Stumps 'R Us. They used TDs with a rubber ring that expands to fill one of the ball's holes and then contracts at the end of the swing. The group rolled strikes and spares while I kept the gutters busy.

We cross the hallway into the din of the conference breakfast buffet. Chafing dishes and mini muffins. Someone passes a leg across a table. Frech and I talk while Judy joins the buffet

---

* I was using the word *stump* until Judy took me aside to explain that some amputees don't like the term. "They say, 'I'm not a tree,'" she explained. Has there ever been confusion? Possibly so! At least two organizations use different stylings of the name Stumps 'R Us: the Bay Area amputee group and a stump-grinding and tree service company in Ontario, Canada (Stumps 'R' Us).

line. He says he encounters veterans and accident victims who've endured ten or twenty years of surgeries on the false hope that one day their limb will completely recover. "They fight it so long that they lose these chunks of their lives, time that they could have spent acclimating to their prosthetic, getting their coordination back, their body back."

Ezra Frech, Clayton's son, comes over to reconnoiter with his dad. He holds a Starbucks bag in his left hand between his index finger and his thumb, which is a toe. Ezra was born with only one finger on that hand, a missing fibula, and a deformed knee. As a child, his leg was amputated mid-thigh and replaced with a prosthesis. Rather than discard the big toe, the surgeon used it to create a thumb. You can go on the internet and watch Ezra Frech win gold in the high jump and the 100 m sprint at the 2024 Paralympic Games. It shames me to tell you what I said to Ezra when we were introduced. I didn't yet know about his athletic achievements—though the Team USA T-shirt should have been a clue—and I asked him, "Can you run?" The assumption behind the question was ignorant and insulting. It was akin to someone hearing me say I'm a writer and asking me if my work has been published anywhere. (Ezra, unlike me, showed no hint of annoyance.)

It is a common and misguided assumption: If you're missing a limb, then you must be operating at a sizable disadvantage compared to anyone with the full complement. There is a strong bias for wholeness, even when, as in Judy's case, wholeness is itself the disadvantage. " 'It still has healthy tissue,' is what they said to me," she recalls. She unpapers a muffin. "I'd go, 'Yes, but I can't walk on it.' " She describes sitting at the front of Walmart, waiting for her family and watching people walk past. "So many of them don't have a normal gait," she says. "They have back pain, or a bad hip, or they're obese." Or old. We will all, should we live long enough, be disabled.

The irony of applying the word *disabled* to Ezra Frech is

especially rich. Compared to Ezra Frech, I'm disabled. I don't have a prosthesis,* but that's just because there's no prosthesis for sixty-four and too lazy to get in shape. By using sports prostheses and TDs, Ezra Frech can, I would wager, do pretty much any activity an able-bodied person can. Or as he says, "do it a little differently." Ezra uses a rock-climbing foot, and has blades for running, basketball, and soccer. There are surfing feet, swim fins, pivots for kayaking, handlebar adapters for mountain bikes. TDs exist for archery, lacrosse, table tennis, basketball, fishing, golf, ice hockey, softball, baseball, uneven bars, martial arts, billiards, volleyball, bowling, weightlifting, and cricket. Musicians can buy adapters that grip drum sticks, violin bows, guitar picks. Should no TD exist, there is a maker space whose engineers can create something for you. Most recently they designed an adaptive device for folding origami. Judy knows a prosthetist who made a mop foot for his wife.

Even a century ago, a prosthetic leg seemed barely to hold one back. In the 1921 edition of *Handbook for the Limbless*,† a World War I vet identified as W.B.B. describes his "Saturday on an artificial leg." After swimming and diving at the local baths, he bicycles five miles to a cricket match, where he takes four wickets and makes plans to attend a dance. The handbook offers tips for horseback riding, billiards, fishing, and boxing. ("For a man who has lost an eye I say take out the artificial eye when boxing.") Another vet describes sporting activities at the Brighton Pavilion Hospital for Limbless Soldiers: bowling, swimming, tennis, and football. "Stoolball was also played, but in

---

* Oh wait, I do. Without my artificial lenses, I'm legally blind. Ezra puts on a leg in the morning; I put on contacts.

† The term *limbless* suggests that this was an instructional guide for quadruple amputees (which would make *Handbook* something of a cruel joke). In odd fact, *limbless* was a post–World War I euphemism in use in England for anyone missing even one limb.

comparison to other games it did not prove very popular." Different name might have helped.

Given all this, it's easy to understand why an amputee might prefer the term "differently abled." Ezra, though, has no issue with the word *disabled*. What bothers him is the hypocrisy of social media trolls who call him disabled at the same time that they accuse him of having a competitive advantage—of "techno-doping"—because of his blade.

"You mean like Oscar Pistorius—like he wouldn't have won otherwise?"

"He didn't win," Ezra says. "He made it to the semifinals and that was it." It was the second time an amputee competed in the Olympics, and no one has done so since. "Think about how many people are amputees," Ezra continues. "If it were an advantage, there'd be more than one guy." Ezra is, Judy told me later, the first above-knee amputee ever to receive a full athletic scholarship to join a Division I track team alongside able-bodied athletes.

Ezra competes to win, but more importantly, to him, he competes as a way of challenging stale conceptions of what it means to be disabled. He likes to think about the number of limb-different people he's inspired through his TikTok videos or by walking around in shorts.

"Ezra never wears pants," his dad volunteers. (On a more practical plane, people with prosthetic legs often prefer shorts to pants, because you can't point your toes very far with most prosthetic feet. Judy Berna explains: "To put on my pants, I take off my leg.")

We say so long to the two Frechs and head into the exhibitor space in the Sinatra Ballroom.* Toward the end of one row

---

* Frank's last performance took place in this very room, in February of 1995. But do not think that Frank Sinatra ended his mighty career with a gig at a Marriott. Back then, this was a tony resort that hosted a prominent celebrity golf tournament.

of booths, an intense-looking man with a shaved head gives us what I've taken to calling the prosthetist's glance: As they're saying hello, their gaze dips to your legs and quickly back. Just to see: Above knee or below? What are you using? The prosthetist is Kevin Carroll. He has definite feelings about limb salvage and elective amputation, especially in the case of a child. "This week I was working with a kid who'd been in a horrific accident," he says, the peaks and troughs of an Irish accent making trauma sound like a Joyce reading. "Here's the surgeon patchin' things back together, bit by bit, year after year. And now he's seventeen. Still strugglin'. Versus if they'd done it when he's a little kid, he'd be up and literally running around after four, five months."

Amputation that's not medically necessary is of course a fraught scenario for a parent. "There's a hope that something will be developed one day in the future," Carroll says. "Or they may say, 'Oh, we're going to wait till she grows up and let her make her own decision.' The problem is that by then, psychologically, the defective limb has been incorporated into their identity. Often they never get it done."

Carroll feels surgeons supply some of the false hope, either directly or indirectly, through their hesitation. "Their job is to save," he says. Lives, limbs, their reputations. Amputation may be perceived as failure. Or grounds for a lawsuit. "It's a little scary from a liability standpoint," allows Andrew Rosenbaum, a foot and ankle surgeon I spoke with after I got home. "You can't be faulted for just whittling away, cleaning it out, prescribing more antibiotics." Rosenbaum, who teaches at Albany Medical College, had amputated a lower leg the day before. "You try to guide them to the decision on their own. I present it as another reconstructive option, rather than a last resort." As ever, insurance plays a role. Before insurers will reimburse for the cost of an amputation, a case must be made that it's warranted.

I ran all this by one of Judy's own surgeons, the Fort Collins, Colorado, reconstructive foot and ankle surgeon Wes Jackson.

He thought the hesitation likely had more to do with the surgeon's expertise and experience. "The less you do of a procedure," he said when we spoke by phone, "the less comfortable you are with managing it and taking care of it."

Amputation is not the quick and simple operation it once was. Gone, mostly,* are the days of the "guillotine amputation": a few strokes with a curved blade and an amputation saw, followed by the press of hot irons to stanch the bleeding. (Before the advent of anesthesia, speed was understandably as important as skill.) In centuries past, technical advances took the form of, say, an amputation saw with fewer decorative frills, because the frills harbored bacteria and would, to quote London's Wellcome Collection catalogue, "catch on soft tissue." Or a saw in the form of a chain of serrated links† to be threaded around the bone to cut it from behind, after the flesh in front is cut with an amputation

---

* Emergency personnel who need to free an accident victim trapped in a vehicle or a piece of industrial machinery sometimes lack amputation equipment and must use what they have on hand—often a reciprocating saw (Sawzall) or a Holmatro rescue cutter (normally used on the car). To determine which is least damaging (to both victim and rescuer), UK researchers Caroline Leech and Keith Porter recruited firefighters and cadavers. Though both tools achieved the goal of speed—the leg being freed in under ninety-one seconds—no obvious winner emerged. The Sawzall caused not just "significant spray/splattering of blood" but outright aerosolization of tissue, requiring rescuers to wear masks. The Holmatro injured an extra two inches of soft tissue and had the purely aesthetic but not to be dismissed downside of a "loud splintering sound." The firefighters were contacted a month later to check for lingering psychological reactions.

† Department of ghastly but true sentences: The chain saw got its start in the delivery room. It was a manual tool, developed by Scottish obstetricians in the late 1700s to temporarily open up a small pelvis when the baby's head was too big to pass through (and Caesareans often ended in death). The flexible chain could be threaded around the bone, so it could be cut from behind and without slicing through the soft bits. Surgeons today sometimes cut bone with a similar flexible blade, called the Gigli saw, after the Italian obstetrician who designed it. The surgeon who told me about it pronounced it "giggly saw," and I like to think that's how he asks for it in the OR. While we're on the topic of surgeons whose names clash with their titular inventions: the Rust Amputation Saw.

knife. Today, amputation is a half-hour operation that often involves measures to prevent phantom limb pain. I watched a lower leg amputation in which the surgeon would take the end of any major severed nerve and embed it in some muscle, to "give it a task." So that it's not, to use another surgeon's metaphor, a loose power line sparking in the roadway. The technical term for what I observed is "regenerative peripheral nerve interface." (The surgeon—it was Jeremy Goverman, from chapter 2—called it a "muscle burrito.") Providing pain-obliterating or nerve-quieting medication in the lead-up to the surgery, as Judy's surgeon did, is also thought to help.

By the reckonings of more than a few journal articles, the decision to amputate should be fairly clear-cut. In study after study, when researchers compare patients who'd had below-knee amputations with a similar number who'd had limb salvage surgery, the amputees end up with better health and "functionality" and lower pain scores. "Amputation should be strongly considered when confronted with a borderline salvageable tibial injury," concluded a group out of New Zealand. Kaj Johansen and his coauthors, in a 1990 *Journal of Trauma* paper, wrote with a blatancy not often evident in surgical journals: "It has become increasingly clear that a population of trauma victims exists in whom, because of a combination of surgical enthusiasm, clinical misjudgment, or wishful thinking, limb salvage has been attempted when, in fact, such efforts are doomed."

Scoring systems[*] exist to help distinguish limbs worth saving from those that aren't, but the decision ultimately comes down to the surgeon's judgment. It's easy to imagine how patients' hopes and fears, and those of their families, however misguided, might erode a surgeon's conviction.

A few months before this conference, lingering infection

---

[*] For example, the Mangled Extremity Severity Score, which I here single out for the dark humor of its acronym.

developed in Judy's remaining biological foot and she landed in the hospital with sepsis. This time around, it was the surgeon who brought up amputation. One reason she's come to the conference is to shop for new feet (plural because she'll need a matching pair). I'm tagging along because I'm curious about the state of the art. The better prosthetics become, the easier it will be to make a case for elective amputation.

Feet are sold separately. I had not known this. A prosthetic foot (and ankle) attaches to a prosthetic leg, and over the foot goes a "foot shell," to fill out a shoe or to make the whole appear foot-like. Shells come in a palette of skin tones, and if you like you can buy a pair with a space between two toes, for wearing sandals. A foot may incorporate something to help propel the leg while walking—hydraulics or a small C-shaped "blade" under the heel.

The vendors display their products on tabletop pedestals and in Windexed vitrines. A few of the feet are posed on artificial boulders to highlight how they tilt from side to side, which helps on uneven ground—rugged trails and other "challenging terrains of adventure." Sales materials for prosthetic feet are a little like those SUV ads you see on TV, with fleece-clad drivers four-wheeling over boulders and scree to summit a mountain ridgeline. When really, the average driver has no interest in bringing the car up onto a ridgeline. He just wants to get over to Costco before they close.

Judy is a follower of "biomechatronics" visionary and MIT professor (and double below-knee amputee) Hugh Herr. His research, through the Center for Restorative and Regenerative Medicine and elsewhere, has given rise to "smart" prosthetics that incorporate microprocessors and sensors to adapt a prosthetic to an individual wearer's gait. Judy was excited to have been a beta tester for a Herr foot, around 2016. While impressive, it seemed fragile to her, and whether or not it was, that made her uncomfortable. "I'm constantly knocking my leg into things, banging it around getting in and out of the car. I need something I can't destroy." Limbs with motors and microprocessors may not be waterproof. You need to recharge them. They're heavier* and of course more expensive than conventional prosthetics. In the words of a prosthetist I later sat next to at lunch, "Everything is a trade-off."

For someone with an above-knee amputation, the trade-off is likely worth it. These limbs involve two joints, not just one. And the taller the prosthesis, the less stable the wearer. Among other benefits, a "smart" leg will sense when someone is losing balance and lock up to prevent a fall. I'm seeing quite a few of these legs today. The electronics are housed midway up the lower leg, giving the wearer a sort of high-tech calf muscle.

What I'm not seeing are those cyborgy hands with multi-articulated fingers. I haven't seen anyone wearing one, and just one company booth. The sales rep has set up a poster with a studio-lit blowup of the hand, with a raspberry held between its thumb and forefinger. Judy laughed as we passed by. "Are you going to spend fifteen seconds picking up a raspberry? No. You're going to reach over and do it with your other hand."

The sentiment was echoed in a 2022 essay entitled "I Have One of the Most Advanced Prosthetic Arms in the World—and I Hate It." The writer, Britt Young, was born without her left forearm. Inspired by a news piece, she had splurged on a

---

* Though probably not as heavy as yesteryear's wooden legs. Until around 1920, a belief persisted that in order for an amputee to achieve a balanced walk, the weight of the prosthesis should match the weight of the remaining leg. World War I veterans in England often preferred crutches to the discomfort of a twenty-five-pound leg suspended from a network of shoulder harnesses. The 1921 decision by His Majesty's Minister of Pensions to make available the new light metal limb was met with joy. "I can dance the whole evening with the light leg," enthused one vet quoted on the matter in *Handbook for the Limbless*.

myoelectric arm with a $72,000 price tag, only some of which her insurance paid for. At an "arm party" the week it arrived, she used it to cut the cake. "And that," she writes, "was one of the last times I ever used it."

Myoelectric limbs like Young's have sensors in the socket that detect electric impulses from muscles in the residual limb. On hers, a double flex toggles the device through a menu of grip patterns, including one for, appropriately, handing over a credit card. "The 'power grip' (a fist) is cool," she writes, "but, oh my God, trying to cycle through all the modes to find it is infuriating." Plus the arm felt "heavy as hell." Everything about it, she says, was exhausting. "Prosthetic arm technology is still so limited that I become *more disabled* when I wear one."

More than half of arm amputees end up abandoning their prosthesis—a figure that has remained steady in the age of myoelectric options. Many never use any kind of prosthetic limb. Instead they've become skilled at doing things with one hand (and a mouth and a torso and two knees). It made me think of my dad, who by his late seventies had lost his molars. Rather than get dentures, he gamely beavered (my brother's wording) his food with his frontmost teeth. Which never slipped or rankled his gums. He gnawed lamb chop bones and ate corn on the cob without a thought.

Later that afternoon I attended a session for arm amputees that included a slide presentation about resources for day-to-day life. The presenter shared YouTube channels and life hacks—helpful devices designed for other uses: a carpet drier set up outside the shower stall ("high velocity but works well"), an operating room foot-pump soap dispenser. I sat behind a man with two body-powered prehensor, or "claw," prostheses controlled by cables attached to an upper body harness. Every now and then, the man would reach up and quickly adjust his COVID mask. When he spoke to the group, he gestured with one or both of his prehensors. They appeared to be as much a part of his physical self as my hands are for me.

Judy and I are sitting in the hallway watching Ezra Frech give running pointers to a group of above-knee amputees. She is talking about media coverage of prosthetic limbs, and how she wishes they'd dial back the hype. "I can't tell you," she is saying, "how often friends send me news stories about the 'next big thing.' I try to be polite. But no, there will not ever be a twenty-dollar custom 3D printed leg that means you don't have to go to a prosthetist."

A word about prosthetists. They are not, or are not primarily, creators of prosthetic limb technology. A prosthetist's bread and butter is sockets: making them, fitting them, adjusting them. The socket is the cup, or "bucket," into which one slides one's residual limb. A socket will need to be adjusted as the muscles of the residual limb atrophy, and if the wearer gains or loses weight. Since the contours of a residual limb are irregular, a socket that fits snugly in one area may be loose elsewhere. Loose means rubbing, and rubbing means irritation. Blisters, ingrown hairs. Prosthetists tweak the socket to remedy these things. They soften the plastic and bump out the spots where it's digging or chafing. They also keep an eye on the person's gait. A good prosthetist, Judy says, can tell what needs adjusting just by watching her walk down the hall.

I'm embarrassed to admit that my knowledge of residual limb sockets comes from cartoon images of pirates—the wood bucket above the "peg." As though the end of the residual limb simply rested on the bottom and took all the weight. That, I now know, would be a disaster. The human foot is cushioned by a layer of sturdy fibrous padding at the heel and ball. A residual limb lacks that. A leg socket bears the person's weight by compressing the thigh (or calf). Think of walking downhill in hiking boots. If they're properly fitted and snugly laced, the whole boot supports the hiker's weight. If they're loose and the toes slide down and take weight at the end of the boot, that will be a painful hike.

An entire industry exists around socket fit. Judy and I have landed at the Comfort Products booth. The proprietor, Fred Lanier, is an old-school sales guy—table skirts and price lists in plastic sleeves. Fred talks us through the product line: cushioning socks in single-, double-, and triple-ply. Lotions to reduce friction. He holds open a sock. "It's lubricated!" he exults. "Stick your hand in there." Judy moves away. The sock swings over to me. "Feel that? It's got mineral gel built right into the fabric."

Two men standing alongside us are examining an antiperspirant for residual limbs. "I have a friend who works for the Department of Transportation, out on the roads," one is saying to the other. "On hot days, he'll pour a half cup of perspiration out of his socket."

What if you could skip the flesh-squeezing, perspiration-collecting socket altogether and just have the prosthesis attached directly to your bone? Orthopedic surgeons routinely hammer and screw metal supports into bones. Why can't you do that with a prosthetic leg? You can, in fact. Judy directs my gaze to a group of amputees milling around the booth of the Osseointegration Group of Australia (OGA). "A lot of the chatter in the amputee world right now is about that surgery," she says. "About whether it's a good option." We head over there.

With a bone-anchored, or osseointegrated, prosthesis, the skeleton bears the body's weight. That makes it feel lighter and more comfortable. And unlike a socketed limb, it delivers sensation, through the skeleton's impact with the ground. This "osseoperception" lets one know by feel what type of terrain one is walking on, grass or concrete or someone else's foot. That is one thing, possibly the only thing, that Ezra Frech can't do. "I am always," he mentioned earlier, "looking down to be sure I'm not stepping on someone." In short, an osseointegrated prosthetic leg feels and moves a lot like a meat leg. It's the difference between dental implants and old-school dentures. In fact, osseointegration—the technique and the term—was pioneered by an oral surgeon, Per-Ingvar Brånemark, "the father of dental implantology." Dental implants—artificial teeth anchored in the jawbone—have eclipsed old-school dental plates, the slipping, gapping, food-grouted uppers and lowers one Poligripped to one's gums. Today I've seen maybe four people with osseointegrated prosthetic legs. What's holding things up? Why is everyone still walking on dentures?

Part of the display at the OGA booth is a satisfied customer in shorts. A few feet back from the booth, Judy stops in her tracks. To stare. We're staring. "I find that very . . ." You can tell she's sifting through adjectives. "Startling."

It's the suddenness of the transition. There is the man's thigh, the pale, soft loaf, and sticking out of it is a metal rod. I'm trying not to take you in the direction my head has gone, the direction of industrial accidents and impalings. Picture instead a flamingo: the broomstick legs supporting the plump, round body.

The man on display, Glenn Bedwell, is an above-knee amputee with a jolly air and a firm handshake. He was "socket-based" for twenty-one years. He was fine with it, despite minor drawbacks—ingrown hairs, perspiration, an inability to sit comfortably on a barstool. Then his skin stopped tolerating the silicone of the suction sleeve he was using to secure his prosthesis and he made the move to osseointegration. He says he can now walk ten miles comfortably.

"People who've had success with this say it's hard to even describe how much better the control is," Judy says. An acquaintance of hers, a double above-knee amputee, made a temporary move to Australia to have it done. For her, it didn't end as nicely as it has for Glenn. Seven years later, after multiple infections and surgical revisions, she's still using a wheelchair.

"It's really taking off," an American surgeon at the OGA booth is saying. He holds his hand at a karate-chop slant. "The curve is like this." Steep. He says this has to do with the FDA's December 2020 approval of osseointegration for above-knee amputees. The system the FDA approved is on display at a booth on the far side of the room. It's the OPRA Implant System, made by Integrum, the company founded by Rickard Brånemark, Per-Ingvar's son.

Integrum also has a happy patient on hand, and a plastic model of an osseointegrated thigh, sliced like an Easter ham, to reveal the various layers. With Integrum's system, the hardware is installed in two operations, three to six months apart. (The Australian process is a single surgery, with patients done and on their feet in, the brochure states, three to six weeks.) In the first operation, screw threads are made on the inside of the bone, and the abutment—the rod to which the prosthetic limb will attach—is screwed into place. The wound is closed, and the bone is left alone for six months to grow into the rod. In the second operation, a plastic surgeon seals the exit point, somehow persuading the skin to fuse with the end of the bone.

The Australian procedure leaves the opening incompletely sealed. Glenn showed us a padded circular collar around his abutment, a sort of residual limb mini-pad. "I still dribble some," he allowed. Though he hasn't had an infection in years, he had six or eight in the first year. One was serious enough to require intravenous antibiotics. Hopefully it wasn't as severe as what befell the patient profiled in a *Sydney Morning Herald* exposé. The patient, a former paratrooper, told reporters he endured years of oozing pus and blood and a stench so intense he claimed he could taste it. He says he was told to spray the leg with Febreze. He compared the pain to "a welder blowing on my legs" (the torch, presumably, not the person).

Integrum's Brånemark made the move to limb prostheses partly because of how few infections his father had been seeing with the titanium screws he was sinking into jaws. But the mouth, bacteriological cesspool that it is, may have evolved better defenses than the rest of the body. It's also possible that the cesspool's harmless resident bacteria are able to outcompete harmful interlopers. One surgeon actually tried transplanting gum tissue around the abutment where the post exits the body, in hopes that infection could in that way be prevented. Others have studied antlers—a trouble-free example of bone protruding through skin.

Whatever the reasons, limbs are not as accommodating as gums and elk skulls. In the most recent OPRA Implant System study, which followed fifty-one patients over a span of ten years, 29 percent of them had a deep (bone) infection. However, a 2024 review of twenty osseointegration studies (of varying durations) found just over 3 percent of patients had had a deep infection. For someone like Glenn, who'd been having significant problems with his socket system, it may be a risk worth taking.

Though those studies all looked at legs, osseointegration shows perhaps the most promise with arms. A new technique called targeted muscle reinnervation rewires the nerves of the arm to amplify the signal from the brain. This allows a person to communicate more effectively with some of the newer myoelectric arms, enabling faster, more fluid finger movements. But as Britt Young mentioned, myoelectric prostheses are heavy—up to four pounds; integrating them directly into the bone, rather than using a harness, would take some of the weight off the upper body.

"We integrate the rewired nervous system into the post," said the Johns Hopkins plastic surgeon Rick Redett, when I spoke to him at the outset of my research for this book. By "post" he means the metal osseointegration rod inside the bone. The electrical signals from the nervous system would travel through the post and into the computer inside the prosthesis.

When I checked back, two years later, Redett referred me to his colleague Jaimie Shores, who had left Hopkins for the Orthopedic Center of St. Louis. Shores reported that only a few patients, nationwide, had received an osseointegrated myoeletric arm prosthesis combined with targeted muscle reinnervation. Integrum was hoping to begin a trial with a few more.

Judy Berna is not tempted by osseointegration. Having recently spent weeks with an antibiotic IV in her arm, fearing sepsis and feeling nauseous, even a small risk of infection is a deal-breaker.

I caught up with Judy a few months after the conference. She'd recovered from her second amputation and had just been fitted with her new legs. She described them as basic: functional and comfortable, which is everything she wanted. She sounded great. She sent me a short video of herself on the sidewalk out in front of her house, holding hands with her one-year-old grandson, both of them newly up on their feet, getting their walk on.

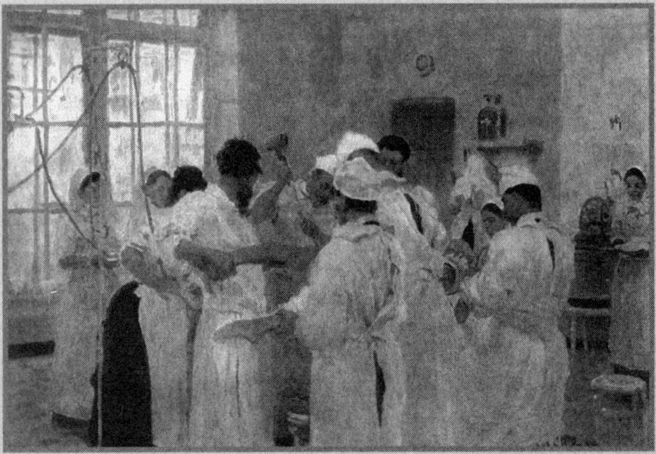

8

Joint Ventures

Woodworking Without Wood

The third hip replacement of the morning looks very much like the second and first. The patient and the whole operating table are covered by surgical drapes, resembling not so much a person having surgery as a small vehicle under a tarp. A surgeon stands alongside, holding a metal instrument in a hole in the patient's side. The hole—the incision—is held open by a circular plastic retractor the size of an automobile gas cap. From where I stand, six feet back, this is all I can see. Hip replacement has the visual drama of a visit to a Chevron station.

It's the sounds that undo you. The whine of the bone saw as the surgeon cuts off the knob at the top of the thigh bone. This knob, the "femoral head," is the "ball" of the ball-and-socket that makes up the human hip joint. The socket, the acetabulum of the pelvis, sings a sound of its own: the tires-on-pavement squeal of the reamer as it grates away damaged cartilage and resurfaces the bone, preparing it for its cup-shaped replacement. Next, the chattery whir of a rotary drill sinking a screw, this one securing the socket cup while the bone grows into it.

Anyone with a basement workshop is familiar with these sounds, these tools. According to the Wisconsin orthopedic surgeon Paul Anderson, at least seven common woodworking joints have also been used in orthopedics. Anderson used to bring surgical residents to his shop to teach them how to craft dovetail joints, which hold fast without glue. To be sure, there are similarities between an artificial hip and a handcrafted cabinet. The supports must be strong and solidly set, at precisely the right angle. Moving parts must fit snugly yet glide evenly and smoothly.

It isn't the same, though. A cabinet has no immune system. It doesn't throw up defenses against building materials it perceives as hostile invaders. It doesn't die under siege from bacteria that gained a foothold on a piece of inlaid metal or plastic. In other words, the surgeon's skill can take you only so far. It's the materials guys you're depending on for a lasting, complication-free build.

"This is an Actis stem, with a Pinnacle acetabular cup." Mike Olmes is a materials guy, an implant rep for DePuy Orthopaedics. Joint replacement is a hardware-heavy endeavor. For reasons practical and financial, it is commonplace today for reps to be on hand in the OR. With the aid of the real-time X-ray movie now playing on the screen of a mobile fluoroscope, Mike will help select the proper size implants and guide their placement. He will not, no matter how hairy things may get, step in and take over for the surgeon, as some sales reps did in the 1970s. (Reps for one medical supply company, its president admitted to the Medical Practice Task Force of New York State, had scrubbed in and taken over a portion of the surgery on 3,240 occasions.)

We're inside an operating room at the Center for Joint Replacement, part of Washington Hospital Healthcare System in Fremont, California. The surgeon is Alexander Sah, of Sah Orthopaedic Associates—*Sah* pronounced just as you'd hope, for an orthopedic surgeon. By his tally, Sah does more joint replacements than any other surgeon in Northern California. Today he'll do ten: hips in this OR, knees in the OR across the hall, crossing back and forth as the day unfolds.

Mike and I are down by the patient's foot, which is Velcroed into a boot at the end of a leg holder, a long rod that Mike moves around by hand. When the thigh bone is no longer connected to the hip bone, a leg defies the constraints of human anatomy. Once the femoral knob, or head, has been removed, the foot rotates freely, owlishly, almost 180 degrees. This creates better access for the surgeon and an arresting visual for the visitor. "Last week a patient asked me what the boot was for," says the anesthesiologist working the OR today, Kris Kuhl. "I was like, 'Lady, you don't want to know.'"

The first total hip replacement took place in 1938. Both the stem and the socket cup were of stainless steel. These days titanium is the metal commonly used, because bone grows into it so nicely and so quickly. When bone and implant fuse, there's no movement going on in there. Movement leads to loosening—and pain and sometimes complete failure of the new joint. The ball on the end of this particular femoral stem, however, is not titanium. "Metal-on-metal" has a troubling history. As a patient walks and the ball swivels in its socket, the friction between the two surfaces creates bits of debris. Depending on the size, number, and makeup of these fragments, "wear debris" can prompt an inflammatory reaction that destroys soft tissue surrounding the joint. I seem to recall an especially messy, high-profile metal-on-metal recall.

"That was my company," Mike volunteers. "We were doing fifty percent of our patients metal-on-metal." Specifically, it was a cobalt-chrome alloy. Mike is built big, college football–big, with a friendly, welcoming demeanor. He's talky, in the way of sales reps, but also straight-shooting, very much less in the way of sales reps. It's possible he doesn't know why I'm here. Johnson & Johnson, which owns DePuy, settled thousands of lawsuits over their ASR (Articular Surface Replacement) prosthetic hip, at a total cost of around $4 billion. "We were really burned by that," Mike says. As were, it must be added, a lot of patients. After two years, the ASR had failed in 5 percent of those who'd received it. A failure rate of more than 2 or 3 percent, Mike adds, is considered within the industry to be "catastrophic."

My attention is yanked from the conversation by what an Actis copywriter describes as "moderate strikes": Sah with his surgical mallet, sinking the implant stem into the core of the femur. I know this sound from family camping trips. It's the sound of a tent stake going down. *MIKE! OH MY GOD, MIKE!* I am saying with my face. And then, with my mouth: "Don't people kind of need their bone marrow?" Mike replies that the implant doesn't destroy the marrow, mostly just compacts it.

Mike studies the fluoroscope images and calls out numbers. Based on these, Sah makes adjustments to the implant's position, helping ensure that the patient's legs end up the same length. And, later, when the cup is installed, ensuring there are no hot spots that could create friction and wear. These sounds, too, are familiar—the *tap-tapping* of a sculptor with a chisel and mallet.

The last major step before closing things up is to cap the stem with a ball. Picture, if you've been around that long, a stick shift knob on an old Jeep. This one is made of a ceramic that's chemically strengthened. Earlier ceramic knobs would sometimes crack or shatter when patients fell or otherwise heavily smacked their hips. Alumina ceramic, as this is called, is outstandingly hard and sturdy, with a very low coefficient of friction. The brakes on Formula One race cars are alumina ceramic. "There's very little wear, provided you install it at the right angle," Mike says.

Another reason, historically, to get the angle right: mispositioned ceramic components can squeak. In a 2009 paper entitled "The Squeaking Hip," 33 percent of surveyed patients with ceramic-on-ceramic hip replacements reported hearing noises while they walked or climbed stairs or bent over. Two patients experienced it when putting on pants. One said it happened during sex. How loud are we talking about? Certainly not as loud as the noise made by a 1950s acrylic implant that—one patient believed (and one author kinda doubts)—was the reason his wife "avoided being in the same room as him whenever possible."

Sah is about done here. He steps back to let an assistant close the incision, thanks the team, and turns to go. Knee replacement number two is prepped and ready in the OR across the hall. He'll be back soon. Sah's "cut time" for knees is just under an hour "skin to skin"—the time elapsed between "skin open" and "skin closed."

One of the first surgeons to move away from metal-on-metal was the English orthopedist John Charnley, back in the 1950s. While Sah is off with his knee, let's take a moment for John Charnley, the surgeon who did more, and occasionally less, for complication-free hip replacement than any other. Early on, Charnley pinpointed friction as the pressing problem to be dealt with. He had a dream of an artificial hip joint with the same low-friction glide as human bone on cartilage.

Sometime in 1956, an acquaintance of Charnley's in the Lancashire chemicals industry told him about a promising new low-friction polymer: polytetrafluoroethylene, later and more widely known as Teflon. In its non-cookware state, Teflon is a white, carvable, semi-translucent plastic. It even looks like cartilage. Intrigued, Charnley brought some to his lab. He ran some friction tests, and he liked what he saw. Up in his attic, where he had an office and a workshop, he and his technician, Harry Craven, began fashioning Teflon hip sockets. They'd craft a socket by night, and by day Charnley would put it in a patient, combining it with a stainless steel stem already on the market.

Charnley was a man of singular focus. Orthopedics seeped into all that he did. The front of the family's 1946 Christmas card is an etching, by Charnley, of a surgeon holding a power tool to a femur, rays shooting off from the point of contact in a design that suggests—I think?—the Star of Bethlehem. According to his biographer, Charnley wallpapered a room in the family home with high-power magnifications of the internal architecture of the femur.

Four years and three hundred operations later, it was clear something had gone terribly wrong with Charnley's Teflon sockets. The metal knobs of the patients' femoral implants had worn deep hollows into them, and bits of Teflon debris provoked an angry immune reaction that destroyed adjacent bone and tissue, creating masses of amorphous "cheesy material." The implants loosened. The patients were in pain. Charnley set about removing them. His wife described "an all-pervading gloom." (It had to be tough on her. Even on a good day, this was a man who by his own admission loved his Aston Martin sports car "more intensely than any woman.")

When it comes to replacing body parts, things that make intuitive sense don't always pan out. Bodies are complicated in unpredictable ways. You can't know for certain how a material will perform or react until you put it into a patient and watch what happens, for five, even ten, years. Yet the time required by the FDA to establish the safety of a new medical device is often shorter. And companies can make use of a regulatory loophole—510(k) clearance—wherein clinical trials aren't required if the FDA can be persuaded that the device is "substantially equivalent" to one already on the market. But as the ASR made clear, even minor changes can have a major impact on patients' well-being.

Charnley soon regained his footing. About a year after the Teflon fiasco, Craven heard about a novel low-friction plastic being used to make gears for automated looms. He ran tests. After three weeks, day and night, on Charnley's rig, the material showed less wear than Teflon had after just twenty-four hours. That material was ultra-high-molecular-weight polyethylene, a version of which is the material that lines the titanium hip socket Dr. Sah just installed.

Charnley's other helpful obsession was infection control in the OR. Humans are a constant weather system of wafting, settling bacteria—bacteria drifting along on flakes of shed skin, raining down in droplets of sweat, blowing in on coughs and exhalations. Hip replacements involve large wounds, sizable implants of foreign materials, and strenuous activity on the part of the surgeon, all of which contribute to the risk of bacterial infection. Charnley sought the expertise of the Howorth air systems company—makers of, among other things, air filters to reduce contamination of beer vats. Together they devised a 7-foot by 7-foot ventilated surgical enclosure nicknamed the "greenhouse" for its plate glass sides. Only the patient's body, the surgeon, a nurse, and the leg holder (which used to be a human being) were allowed in. Everything else, including the anesthesiologist and the patient's head, stayed outside. Air was refreshed 100 times an hour. A version of Charnley's clean room—with air pressure-delineated zones instead of physical walls—is used in operating rooms today.

The helmeted "moon suits" worn by Sah and his team today are also descendants of a Charnley invention—the hooded, ventilated surgical gowns he'd developed by the mid-1960s.* In an experiment later published in *The Lancet*, he'd shown that when surgeons exerted themselves and leaned into their work—as hip replacement surgeons must do—bacteria could pass through the weave of their 1960s cotton surgical gowns. This he demonstrated via what he called "the sausage test." Agar, a growth medium, was packed in a plastic tube resembling a sausage casing. The "sausage" was sliced, and the exposed face was pressed to the front of the surgeon's gown right after a surgery. This was then sliced off to create a fresh surface, and the process repeated in different spots. Upon incubating the slices, Charnley found

---

* Though no longer described as "like that of the Ku-Klux Klan"—Charnley's discomfiting comparison. You can't dive into surgical history for long without bumping up against the casual racism of the medical elite. Twice in this chapter alone. Where did the surgeon Austin Moore, in 1952, go to guinea-pig his new Vitallium hip stem? "The colored unit of the Columbia Hospital."

almost half of them growing colonies of bacteria. Modern surgical gowns are of a more tightly woven, less sheddy synthetic material. Because this safer fabric doesn't breathe, current orthopedic surgical garb may, like Charnley's "body exhaust system," include a small head-mounted motor that cools the surgeons and draws away their exhalations.

Dr. Sah is back, starting in on the day's fourth hip. There goes the bone saw. Did I describe it as a whine? It's more an urgent buzz. Either way: unsettling if the bone is yours. Kuhl, the anesthesiologist, says patients sometimes ask to stay awake for the procedure. Technically they could, because they're numbed from the waist down with a spinal epidural, but Kuhl prefers to deep-sedate them. "Because there's a lot of hammering and sawing."

Charnley's biographer wrote that the surgeon hated noise, that he enjoyed music but rarely attended concerts, as invariably someone would cough and the sound would irritate him. Snoring infuriated Charnley, to the extent that he once, in an officers' quarters, drilled a bone screw through the thin partition separating him from his neighbor, "in order to disturb the offender in a way he would most certainly remember."

It strikes me that it wasn't so much the sounds that irritated Charnley as the humans attached to them. Charnley liked who he liked but found humanity on the whole lazy and stupid. He had a family but preferred the company of other orthopedists. Vacationed with them. Worked on Christmas. Despised chitchat. A scene in his biography has him hosting cocktail parties with "a bottle of wine in one hand and a human femur in the other," effectively warding off small talk. For a misanthrope, John Charnley was a surprisingly committed humanitarian. We can only marvel at the lives saved and improved by this intolerant and condescending man.

While Charnley and his colleagues were experimenting with newfangled plastics, a surgeon in Burma was having spectacular results with a material older than the Pleistocene. I learned of this man through fleeting references in two journal articles. I would have read right over him but for the name: Ban Saw. Dr. Ban Saw, orthopedic surgeon!*

Who, it turns out, never existed. The name is an artless spoonerism that slipped past editors of the *Journal of Biomedical Materials Research* and the *Medical Journal Armed Forces India*. The surgeon's name was San Baw, not Ban Saw. Dr. Baw practiced in Mandalay in the middle years of the twentieth century. He had studied at the University of Pennsylvania, where a researcher, Paul B. Magnuson, had been investigating the properties and medical advantages of ivory. Ivory pegs and screws had been used† to secure broken bones as far back as the mid-1800s and the results had been surprisingly successful. Magnuson wanted to know why.

---

* It's not (entirely) true that I choose the people in my books for their names. My first pick for this chapter was not Sah but rather a prominent Mayo Clinic orthopedist—and the guy to whom, I realized with a jolt, I'd lost my virginity. He didn't answer my emails after the first one, though not, I'm pretty sure, because he recognized my name from back then. A friend from high school and I, townies both of us, had gone to a frat party with the aim of choosing someone cute and "getting it out of the way." Boom, done. The man spends more time inside the hips of a typical arthroplasty patient than he did in mine.

† As related in the journal *Orthopaedics & Traumatology*, the idea of using ivory to replace joints is as ancient as Greek myth. The story—as told by two modern Greeks, University of Athens sports scientists Nikitas Nomikos and Chris Yiannakopoulos—goes like this. One day, Zeus's son Tantalus decides to test the omniscience of his father's godly pals. He invites the entire Pantheon to a banquet, slaughters his son Pelops, cooks the body, and serves it as an offering. The gods pass the test, intuiting the menu and refusing to partake—except for the goddess Demeter, who absently tucks in, consuming Pelops's shoulder. Zeus steps in to reassemble his grandson. He collects the untouched meals and throws everything into his special cauldron, returning Pelops to original, prebanquet form—minus the shoulder. Demeter sheepishly commissions "a splendid ivory replacement for Pelops, thereby atoning for her sin and accomplishing the first shoulder arthroplasty." The End.

What he discovered, and Baw later verified, is that ivory is in many ways ideal for the job. It stands up to the forces of compression and bending. It's durable and biologically nonreactive, and its coefficient of friction approaches that within a natural joint. And it was cheap. In 1911, writes orthopedic surgeon Bartek Szostakowski, author of several papers on the ivory hip implants of San Baw, a physician could buy an ivory peg for about the cost of a pound of tub butter. Of course, elephants today are endangered, so no one today is making—or advocating making—implants from their ivory.

Long before endangered species protections were established, surgeons moved away from ivory, opting for titanium and metal alloys like cobalt-chrome. Was it because the cost of ivory had increased as elephant populations dropped? Szostakowski doesn't think so. He thinks it was because the metal alloys were new. He shrugged. Szostakowski lives in Poland, and we were on Zoom. "That's how it always goes." More modern materials are assumed to be better.

In 1957, after completing his studies in the U.S., Baw returned to Burma to set up a practice. The price and availability of the new metal implants put them out of reach for most Burmese patients, but ivory was plentiful and cheap. Elephant herds still roamed the countryside. Baw knew from his days at Penn that ivory performed well inside a bone. He also knew that when one of the hundreds of elephants that hauled timber in the country's teak forests died, the tusks were sold to the public. Might it be possible, he wondered, to use the material for hip implants? (Though Baw's students say he performed some total hip replacements, the carvings on record were stems used to shore up fractured femurs.)

Burma had a long tradition of ivory carving, which was still thriving in Mandalay. Baw began making the rounds of artists' studios, inquiring and getting nowhere. These were men who had apprenticed for years, learning the intricacies of classical religious and courtly statuary. And here comes Dr. Baw, asking them to carve simple knob-topped stems to push inside people's bones, where no one would even see them. It was like trying to hire Georgia O'Keeffe to paint the janitor's closet.

Eventually one artist agreed. I've seen a photograph of the carver, Tin Aung. He sits cross-legged on the floor of his studio in a striped longyi. You can tell it's humid. His forehead gleams, and strands of thinning hair lie pressed to his scalp. Baw and Aung's collaboration was a beguiling convergence: art and science, tradition and technology, old and new. Over a span of twenty years, Baw rebuilt more than three hundred hips using Tin Aung's carvings. Only seven eventually failed. It was an impressive success rate for the time, especially given the broad, high-stress range-of-motion demanded by the nation's ubiquitous squat toilets. Seventy-six percent of Baw's patients, Szostakowski writes, reported being once again able to execute a full squat.

Over the years, there have been efforts to create synthetic ivory—not for hips but for piano keys. These amount, basically, to ivory-hued plastic, Szostakowski said. "We can combine all the ingredients, but we can't re-create the architecture." I guess I'm not surprised. The progress made over two hundred years of materials science is bound to come up short in the face of six million years of evolutionary tinkering.

One winning feature of the old ivory joint implants was their low infection rate. Even the ivory pins and plates that secured broken bones in the 1800s seemed to resist infection. And this was before antibiotics were in wide use. (Today's hip replacement patients are given antibiotics to ward off infection.) Infections, though uncommon, are still the number one reason for needing a revision surgery after a hip replacement.

Baw's ivory implants were far smoother than contemporary models, and this may have helped. The outer surfaces of today's femoral stems and acetabular cups are intentionally rough, to encourage bone to grow into them. But does this also encourage bacteria to grow into them? "They really like the porous material, all the nice little hiding places," says Paul Stoodley, a professor of microbial infection and immunity at the Ohio State University Medical School. Stoodley and his lab mates have shown this by submerging various hip and knee implants in agar infused with fluorescing strains of *Staphylococcus aureus* bacteria and some pig-brain-and-heart broth for them to enjoy. After a few days, the implants were removed from the growth medium and examined. The nooks and crannies of the rough areas glowed like fireflies compared to areas where the surface was smooth.

Implant manufacturers are perhaps putting their trust in a concept called "the race for the surface." It holds that if the patient's bone grows into the surface of the implant before bacteria have a chance to move in, then the bacteria will henceforth be denied access. Stoodley isn't sure about it. "Some evidence suggests it's not that simple." I asked Sah about this by email a few weeks after my visit. Why would you want the surface of an implant to provide lots of nice places for bacteria to grow? "If there's bacteria there, it doesn't matter if the surface is smooth or rough," he answered. "It's equally devastating."

If bacteria manage to gain a foothold on an implant, it can be horrifically difficult to be rid of them. I spent a couple of Zoom hours in Stoodley's laboratory with him and one of his graduate students, Amelia Staats, whose specific area of study is the biofilm that sometimes develops on joint implants. The two walked me through the ugly genesis of biofilm—bacterial slime.

It's not uncommon to have a few bacteria drifting singly— "planktonically"—in our bloodstream. These solo travelers are easily found and destroyed by white blood cells patrolling the blood. Inside a joint, though, ominous things can begin to happen. Bacteria start to organize. They bind easily to proteins in the synovial fluid that lubricates our joints. They cluster together around these proteins, making it harder for immune cells and antibiotics to get to the organisms in the interior of the cluster; as with a school of fish, only the ones on the outside of the group get picked off. Staats showed me a 6,000× magnification time-lapse video of fluorescing green *Staphylococcus aureus* joining together in ever-expanding "aggregates" in a sea of cow synovial fluid. I watched specks become clumps over the space of an hour. It was one of the most terrifying things I've ever seen.

If conditions seem promising, the clumps set up camp. The bacteria cast out adhesions—sticky appendages by which they are able to attach themselves to inorganic surfaces: ceramics, titanium, plastics, pretty much anything one might use to fashion an implant. They send out word: chemical signals to say, *Proliferate here! Colonize the area!* The "replication phase" is underway. The colonies begin to wall themselves off inside a protective gel layer. Now they are biofilm. Now you're in trouble. You've been slimed. Compared to what it takes to eliminate free-swimming planktonic bacteria, knocking back a biofilm infection requires antibiotic dosages orders of magnitude higher.

And then there are "persister cells." These are highly resilient cells within a biofilm, cells that can go dormant if, say, they don't have enough nutrients or oxygen. They'll happily wait out the lean times, impervious to antibiotics, ready to reinfect you when conditions improve. Even high concentrations of antibiotics may not get rid of persister cells. When patients struggle with recurrent infections, these cells are often to blame.

The final phase of a slime community is dispersal. Bacteria, now traveling solo again, are carried along on the currents of the bloodstream. If they're not captured and annihilated— say, because the patient is old and/or the immune system diminished—some of these bacteria may haul out upon the shores of an implant elsewhere in the body. In cases of mass dispersal—the "exodus phase"—blood-borne bacteria can overwhelm the immune system and kill their host through sepsis.

Good Lord. How is anyone still alive? "How do you people leave the house in the morning?" I said to the pair on my laptop screen.

Stoodley doesn't dwell on it. "There are two kinds of microbiologists," he said. "There are the ones who say, 'Bacteria are everywhere! We've got to sterilize everything!'" His wife is one of those. "Then there's the ones who say, 'Bacteria are everywhere! And yet we've survived!'" That's Stoodley. "I'm very cavalier," he said. I am too, though a little less so now. One thing I'm funny about is drinking from Mason jars. You just know there's mouth biofilm lurking between those hard-to-clean screw-top ridges.

Stoodley sounded dismayed. "I like drinking beer out of a Mason jar."

On the topic of mouth biofilm, I had a question. Given that tooth plaque is a biofilm, you would think infections on dental implants would be rampant. Yet the rates are about the same as they are for hips—around 1 percent. This is surprising, given that artificial hips are implanted under stringent conditions of infection control, whereas installing a dental implant in a mouth is, as Stoodley put it, "like putting it in a sewer." Saliva and possibly gum tissue appear to have natural antibiotic properties. Stoodley mentions "commensal" bacteria: local do-gooders that outcompete the harmfuls for food and turf. "There is a thought that we might actually put bacteria into the joint during the procedure, to outcompete *Staph.*" Surgical probiotics! No one, to his knowledge, is yet trying this.

However you can achieve it, prevention is the way to go. Digging out an infected implant that has meshed with the bone is, quoting Bartek Szostakowski—who had spent hours at it the day we spoke—"a mess." Staats says researchers are looking at ways to prevent bacteria from binding to synovial fluid proteins. Better still: don't let them in in the first place. Other than antibiotics, the one thing that has done the most to bring down the infection rate for total hip replacement is—thank you, John Charnley—sterile operating room practices.

Given all that is done to keep bacteria out of a surgical incision—the moon suits, the HEPA filters and ventilation systems, the sterile drapes and skin coverings—how is it that they still occasionally get in? Where, I asked Stoodley, are they coming from? "For one thing," he said, "there's a lot more people coming in and out of the OR now. There are the reps . . ." And the writers, including the one who just this morning couldn't seem to get her surgical bouffant cap on properly. ("Your hair's stickin' out," said a nurse as she brushed past me.)

"Oh, and this is interesting. There's some data to suggest there can be bacteria hanging out in the deeper layers of the skin." So even if you've sterilized the surface of the skin, he's saying, the scalpel could still pick up bacteria farther down into the incision.

I asked Stoodley whether, given all he knows about biofilm and resister cells and bacteria lurking in the deep recesses of your skin, he would ever consider getting an artificial hip.

"I have an artificial hip." Stoodley is in his early sixties. "I probably didn't need one, but I want to run marathons." And because he runs marathons, he opted for . . . metal-on-metal! A polyethylene liner won't stand up to decades of marathons, and he didn't want to have a second operation, to replace it. Also, the newer metal implants are far less likely to generate wear debris. Stoodley has run four marathons on his artificial hip.

It's impressive when you consider that, up through the mid-1980s, hip replacements were done almost entirely on patients in their sixties and seventies—people whose lifespan was unlikely to exceed that of their new hip. Today people's *cats* are having it done. It's like the little Longfellow girl: when it is good, it's very, very good, and when it's bad, it's horrid. Fortunately, seven decades on, it is almost always good.

While Sah finishes another knee across the hall, I visit with the anesthesiologist. Kuhl has just sedated the next hip patient. We're chatting about the latest Netflix zombie series when he glances at a monitor.

"Excuse me." He turns to the patient and reaches under the blue drape erected over her head like a child's blanket fort. He adjusts the tilt of her chin to open the airway. With deadpan nonchalance, he says, "I overshot the propofol a little. She wasn't breathing."

"How are you so calm? This is life-or-death stuff."

Kuhl looks down at the patient, now snoring like a pug. "This is, like, so routine."

Surreally, it is both.

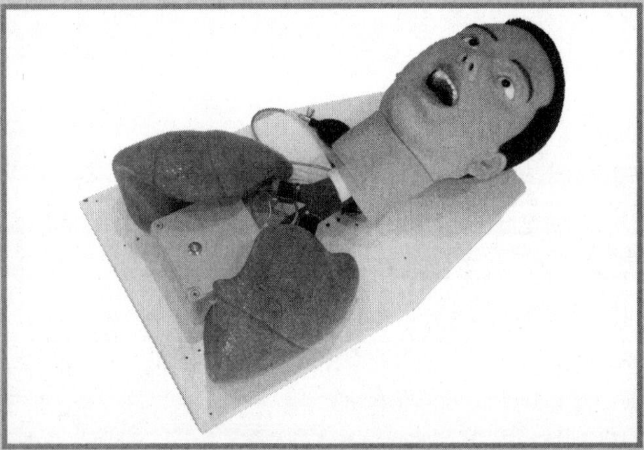

9

Intubation for Dummies

The Brief Terrors of Mechanical Breathing

Until I met one, I thought of anesthesiologists as highly paid security guards. I'd see them sitting by the patient's head, reading, texting, occasionally glancing up at the ventilator readout. What I did not realize is that each of the four or five times I've been permitted to observe surgery, I've been ushered into the operating room after the patient has been intubated—attached to a machine that will breathe for them for the duration of their surgery. Each time, I now know, I'd missed the high-stakes part of the job.

Until I met one, it never occurred to me to wonder why anesthesiologists put surgery patients on a ventilator in the first place. Why would a machine need to breathe for them? It's because general anesthesia has slowed their breathing. And because they've been paralyzed. General anesthesia may include a drug that relaxes (to the point of nonfunction) the muscles. Paralyzing muscles makes them easier for the surgeon to cut through, and it ensures that the patient holds still. If you were to awaken while someone has your guts on a table or your heart in their hand, things could get chaotic fast. Paralyzing the patient also makes it easier for the anesthesiologist to push the breathing tube past the vocal cords during intubation. However, it has to be done quickly, because one of those paralyzed muscles is the diaphragm, the major muscle that works the lungs. Meaning that the patient is on a fast track to suffocation.*

The time pressure is intense. An anesthesiologist has but a few minutes to get the ventilator tube down the windpipe of their poised-to-suffocate patient. The longer it takes, the greater the risks: stroke, heart attack, kidney damage, brain death. All the consequences of oxygen deprivation. Generally speaking, the scariest part of going under the knife is the under, not the knife.

"People don't realize how dangerous anesthesia can be," said Jordan Newmark, who is the anesthesiologist I met. "I've been saying for years, they should make a movie like *Top Gun* but about anesthesiology," he said when we first spoke. At the time, this confused me. It was as though Jordan had access to some bizarro elevator-pitch app that randomly combined hit movies with medical specialties. *Like* Gladiator, *but about urology*. He was insistent: intubation is one of the riskiest maneuvers in modern medicine. "People get surgery willy-nilly, like getting their hair cut. But it's frickin' scary. No one knows."

I wanted to know. Jordan invited me to sit in on a half-day seminar he would be teaching at Stanford University. As part of the requirements, attendees would have to intubate a training dummy. There would be some downtime, so I'd have a chance to learn too.

And so I am here, early on a Wednesday morning, at

---

* One paralytic in the anesthesiologist's arsenal, succinylcholine, leaves no trace in the body, making it a perfect murder weapon, if hard to come by. Detectives solved a famous 1966 double murder case—the killer offed his wife and his mistress's husband within a span of months—in part because the suspect was an anesthesiologist. More recently in the annals of philandering, uxoricidal anesthesiologists, we have Khaw Kim-sun, of Hong Kong. Khaw pumped carbon monoxide into a leaking yoga ball that his wife kept in the trunk of her car. When questioned by authorities, he claimed he'd taken the gas from the hospital where he worked to use on rats at home. But as his housekeeper testified, there were no rats in the house. Or anyway, just the one.

Stanford's Center for Immersive and Simulation-based Learning. The CISL—known to those who use it as "the sizzle"—is one of the first and finest medical sim centers in the United States. An OR, ER, and exam rooms are realistically appointed, down to the pale green paint* and the wall-mounted boxes of latex gloves. We are seated around a table in the CISL debriefing room—myself, Jordan, and a half-dozen residents completing a fellowship in pain management. Jordan refers to them as "the pain fellows," which makes them sound simultaneously jaunty and sadistic.

Jordan has green eyes and brown curly hair cut short, or anyway not long enough to express itself. He wears Hokas with his scrubs and a white lab coat. His right earlobe has side-by-side piercings from a past I can't quite picture him in. He scrawls on a whiteboard as he speaks. Not a word of what he's written is legible. Most things Jordan Newmark does, he does at a fast clip. Writing, speaking. Listening! He plays audiobooks on double speed. He has a lot to do. He teaches, practices, runs a medical legal consulting company, has kids. The desktop of his laptop has forty icons, arranged in rows. It looks like a game

---

* The ubiquity of pale "hospital green" dates to the 1940s heyday of "color therapy," which held that pastels had a calming or cheering effect. Some articles cite the influence of Harry Sherman, who in 1914 had the walls of a San Francisco operating room painted "spinach green." Spinach is dark green, however, and darkness was, for Sherman, the point: "so that the operator who looks up from a wound shall not encounter a glare of light and find his eyes useless for a moment." Had he stopped there, I'd have let it go. "I then decided to surround the whole operation field with black," he wrote in the *California State Journal of Medicine*. Sheets and surgical drapes, the gowns of the nurses and surgeons, the coverings for instrument trays, "were all of black." Staff gently voiced a concern that patients would, upon encountering a team of physicians got up like Horsemen of the Apocalypse, consider it a bad omen. Sherman insisted this was not the case and that his fellow surgeons "appreciated the improved optical conditions." Maybe so, though in time, calmer minds and brighter walls prevailed.

of Concentration. Despite all this, Jordan will reply to a text within minutes. (Often, yes, while sitting by a patient's head.)

On the table across from where I'm sitting is the training manikin that will reveal the high-stakes challenges of intubation. Here's a big one: Humans eat and breathe through the same hole. Everything—lunch, cocktails, coffee, air—begins its journey in the same chute. A few inches along, air must quickly and completely diverge from food and drink: the former heading into the trachea (or windpipe), the latter to the esophagus. The human airway is a railroad switching yard. A flap of tissue called the epiglottis, along with a quick choreography of muscles, executes the switch under orders from the brain stem. For something that feels so effortless, it's frightfully complex. As in a train yard, the consequences of messing up can be dire. For one thing, if food gets into the lungs, it brings with it bacteria, which will thrive in this wet, dark place and possibly bring on pneumonia.

We should have blowholes, Jordan is saying. Whales and dolphins eat with their mouth and breathe through the top of their head.*

In this morning's CISL scenario, paid actors are playing patients with opioid addictions who are making up pain symptoms in order to get a prescription. At the end of it, they'll give the fellows feedback on how well they handled the interaction. The

---

* Why, then, do we see the veterinarians in the navy filmstrip *Man Under the Sea* intubating a dolphin through its mouth? Wouldn't it be easier and safer to intubate the blowhole? It would not, answered cetacean anesthesia expert Shawn Johnson, in an email. Dolphin blowholes, he said, are divided into two tubes with several sinuses, "so it would be like intubating a human through their nose." You'd need a narrow tube, too narrow to deliver enough air for the large lung volume of a dolphin. Also not happening via the blowhole (despite various internet claims): sex. Dolphins breathe exclusively through their blowhole, so a penis would represent a significant choking hazard.

patients are down the hall, waiting in simulated exam rooms. The first group of fellows files out the door. The intubation test, Jordan calls after them, will take place afterward, on the training manikins. "Please don't intubate our actors."

While the rest of the fellows wait for their turn, I get acquainted with one of the training manikins, the Laerdal Airway Management Trainer. He's less a manikin than a bust—a head and shoulders with lungs and a stomach attached. There is something disquieting about his expression. It's meant to convey unconsciousness—mouth partway open, eyes closed—but to me it suggests a little more than that. It makes me think of Bernini's *Ecstasy of Saint Teresa*. And, going forward, the famous work of art will make me, and now you, think of an airway management training dummy.

Step one is to open the patient's mouth and hold it open—with my left hand, because the right hand will be grappling with the tongue. The technique is called the "scissor open." It's the same sort of motion as snapping the fingers, but executed slowly (and snaplessly) and between someone's front teeth. As the middle finger presses down on the lower incisors, the thumb pushes up on the uppers.

While my left hand holds the jaws open, I use my right to pick up the laryngoscope. It's topped with a Macintosh "tongue blade" and resembles a short-handled ice axe. Jordan instructs me to insert the blade into the back of the mouth and use it to scoop the tongue up and off to the side. Otherwise it blocks the view of the airway—and access to the windpipe.

I do as directed. Tongue going nowhere. Jordan gets up from his chair and stands behind me, coaching. "Little bit deeper. Lift up more aggressively. Don't be nice! You're trying to keep this guy alive." Now I'm rummaging more or less blindly, but definitely more aggressively. The dummy begins making a clicking sound, like a dolphin. (If only!) "That means you're pushing on his teeth hard enough to potentially break them," offers one of the fellows, not looking up from his phone. It is not uncommon for teeth and dental work to become collateral damage during an intubation. People often wake up with split lips and sore throats.

I've got the tongue captive now. The blade, which is a couple of inches wide, also serves to prop open the jaw. This frees up my other hand to grab the ventilator tube and push it down the windpipe. If I could only find the windpipe. There's a lot going on down there.

"See that top floppy thing?" Nope. "That's the epiglottis. The vocal cords should be underneath that, and the opening to the trachea is between them." Remember back before GPS and you'd pull into a gas station, lost, and the guy would go, "You're gonna make a left out of the lot and get all the way over to the right. Just past the Arby's, take the slight right, then merge left about a quarter mile past the second light." And you'd get back in the car with your mind a blank. That's me now.

"Do you see the vocal cords? They're white." Nope. "I can see them from here." Not helping.

The easiest intubation Jordan ever did was on a man who had tried to kill himself by slitting his throat. He'd brought the blade into his neck straight on, slicing the windpipe and then backing off before the blade reached the arteries. The man survived, arriving at the emergency room with his windpipe sticking out of his neck. Though he wasn't in a mood to appreciate it, the man had never been breathing easier. Seventy percent of the resistance encountered by airflow is between the nose and the epiglottis. In a breathing emergency, doctors will cut a hole into the trachea midway down the front of the neck so the patient can pull air directly into their windpipe. Tracheotomy: a human blowhole.

Jordan has me tilt the dummy's head, to flatten the curve of the airway and gain a better view. I'm distracted by two fellows beside me, who are rummaging through a box of gadgets Jordan calls the intubation box.

"The Miller blade," one hisses. "I hate the damn Miller blade."

"I like a Mac 3," says the other. "Mac 3, I can do ninety percent of patients."

"You ever play around with the McGRATH? It's a little smaller."

"I *do* like the McGRATH."

Anesthesiologists can get a preview of how difficult an intubation is likely to be by looking at a patient's uvula, the fleshy stalactite at the back of the mouth. Fat deposits in the tongue can swell the organ to the point where it's hard to see the uvula and what lies beyond it. There's a grading system for this, called the Mallampati score: four drawings of wide-open mouths with protruding tongues. The tongues get progressively thicker and the uvulas less and less visible. When my book *Gulp* came out, someone on Twitter accused the cover designer of knocking off the Rolling Stones logo. Not so! He knocked off the Mallampati scale (Class 2—complete visualization of the uvula). A Class 4 tongue completely hides the uvula. That's going to be a trying morning.

I spy an opening and feed the tubing through, fast, like a sailor letting out rope to a drowning shipmate. "Stop," Jordan commands. A few inches along, he explains, the windpipe splits into two branches, each progressing into a lung. If the ventilator tube is pushed beyond this point, as I've just done, it's only going to inflate one lung. There are so very many ways to mess this up.

"I've been doing this for fifteen years and I still get nervous," Jordan says. I try to imagine what's it like to intubate a patient for the first time. Who gets to be the first customer?

"Someone easy," offers a fellow. Skinny, he means. "You'd be prime real estate."

Or not. My jaw doesn't open wider than about an inch and a half. In the UK, there's a club for people like me: the Difficult Airway Society. (A Mitch McConnell chin will also get you in the door.) If you've been difficult, you are added to a registry and given a Difficult Airway Alert Card to carry in your wallet. Any tips for intubating someone like me? The Miller-blade hater looks up from his phone. "I've tried dislocating the jaw. Works really well. You can pop the jaw, just pop it out," he says with maybe more relish than is called for.

"Or you can try this." Jordan holds a wide-mouthed tube that just slides a few inches down the throat and sits on the top of the airway. The danger here is that if you ventilate too forcefully (or the patient's head is angled improperly), some of the air may go down the esophagus and inflate the stomach. That's a dangerous scenario. As a protective mechanism to keep itself from bursting, the stomach will reflexively empty. If the patient is inhaling at that moment, vomit can wind up in the lungs. In addition to harboring bacteria, vomit is partly stomach acid, which can burn and damage the tissue of the lungs.

Should it happen, the anesthesiologist must move quickly to suction the stomach contents from the lungs. Students—though not today's batch—train for this scenario, too. The Laerdal manikin comes with a stomach, seemingly manufactured by the Whoopee Cushion people, and a five-ounce jar of Concentrated Simulated Vomit.* ("Induce vomiting with a squeeze"!)

The same danger applies to emergency bag-masking, the manual version of ventilation. A plastic mask is pressed over the nose and mouth, air pumped in by squeezing a stiff plastic bag. Jordan brought one with him today. He takes me aside. "So we can't take our COVID masks off in here"—CISL protocols—"but if you want to know what it feels like, I can bag mask you, like, upstairs somewhere." It sounds thrillingly indecent. "Or

---

* Like Barbie dolls with their sunglasses and tennis racquets, training manikins have accessories—and none as excessively as Nasco's elderly patient care trainer TERi. The TERi "List of Components" is a depressing excursion into human decline: Hearing Aid, Hospital Gown, Dentures (Upper and Lower), Foot Wounds, Deep Tissue Wound, Edemas (Pitting and Non-Pitting), Enema Bag, Simulated Urine (Quart).

you can try it on yourself." This I do. It makes me need to burp, a sign that I've positioned it wrong, that I'm pumping air into my Whoopee Cushion.

Despite all of this, and because of the skill and presence of mind of anesthesiologists the world over, surgical patients rarely die from intubation-related complications. By one study's reckoning, the rate among ICU patients is 1 percent. Given that surgery patients tend to be in better shape than those who show up in ICUs, presumably the rate among them is lower.

As regards injuries from the forced air itself, these used to be far worse. Forty or fifty years ago, the ventilator was set according to the patient's weight. It turned out that for many people, that was too much pressure and stretch. "The lung is delicate tissue," Jordan says. "People wound up with really stiff lungs." From scarring. "They've dialed it way back."

Ventilators and their potential for lung damage are far more of a concern in the ICU than in the OR, because of how long patients stay on them. Spend more than a few days sedated on a ventilator, and all manner of bad things begin to happen.

Which made me wonder: How did people with paralytic polio survive? These were people who, because their diaphragm and rib muscles were paralyzed, essentially lived in a ventilator—the so-called iron lung—for years. How did they manage? Surprisingly well, in fact. Because the iron lung is a fundamentally different beast.

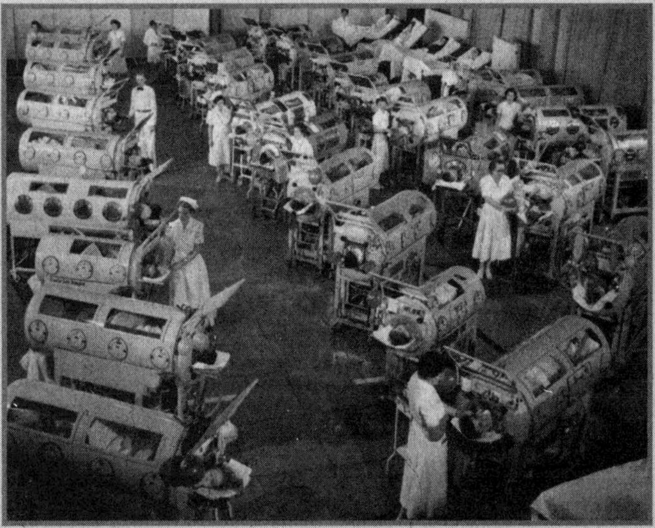

10
Heavy Breathing

Inside the Iron Lung

I don't remember much from fifth-grade science, but I remember the lung. Our teacher, Mrs. Claflin, had told us she had something to show us and disappeared into a supply closet. She returned a moment later with a pig lung, divorced from its partner and encased in a bell jar, which she held before her like a railroad lantern. She showed us how the animal's windpipe passed through the top of the jar and announced she was going to make the lung breathe. I expected her to do this by blowing into it, for I had, somewhat enviously, watched mouth-to-mouth resuscitation being delivered by Marcus Welby's dashing young partner, Dr. Steven Kiley, and other heroic MDs of 1970s television drama.

Instead Mrs. Claflin grasped a pull on the underside of the jar. The flexible rubber base descended, and the lung expanded. She explained that by lowering the air pressure around the lung, she had created a vacuum that sucked it open. As the lung ballooned, the air pressure inside it dropped and outside air was drawn in through the windpipe. Mrs. Claflin had shown us the principle behind human (and pig) respiration.

The medical term is negative-pressure ventilation. Instead of a bell jar contraption, we use the muscles of our diaphragm and rib cage to open our lungs, but the effect is the same. It lowers the air pressure inside them, which causes air to flow in.

Long before Mrs. Claflin, there was "Dr. Lewins of Leith." In 1840, in an Edinburgh hospital, the body of a drowned sailor lay in a sealed box, with only the head protruding. Also protruding from the box was an oversized syringe that created a vacuum by drawing out some of the air. The lowered air pressure in the box caused the dead sailor's lungs to expand and pull in air. The plunger of the syringe was then pushed, the lungs deflated, and the corpse exhaled. "The dead body was made to breathe in such a manner as to lead the bystanders to suppose that the unfortunate individual was restored to life." Dr. Lewins had a showman's bent. "A lighted candle was extinguished by being held under one nostril, and again lighted" when moved aside.* Lewins described the proceedings in a letter to the editor of the *Edinburgh Medical and Surgical Journal*, concluding, rather abruptly, "I have nothing further to add at present, but the reiteration . . . that this invention is one fraught with importance to humanity and medical science."

Dr. Lewins of Leith was correct in that. He had effectively and memorably demonstrated the life-saving potential of machine-generated negative-pressure ventilation. Here was a simple device that breathed for you, if you could no longer do it yourself. Not only did it breathe for you, it breathed *like* you. As opposed to the ventilators of today's medical landscape (and of the last chapter), which use positive pressure, inflating lungs like party balloons.

The ensuing century saw a motley progression of negative-pressure ventilators, most being variations on the dead sailor's box. They were effective for the patient but exhausting for the

---

* The candle must have been incompletely extinguished, as reigniting trick candles did not exist in 1840. The National Candle Association would like them to go back to not existing, because they sometimes relight after they're thrown in the trash and proceed to burn down the house. Canada took a step in that direction in 1977, by banning trick candles, along with sneezing powder, certain kites, lawn darts, and fun.

operator. Until electricity was widely available, around the 1930s, the cyclical push and pull of the syringe or diaphragm or bellows had to be carried out by hand (or foot), making the devices mainly of use for drownings and other brief respiratory outages. You couldn't use them for someone with chronic breathing issues, because who's going to sign on to turn a crank or pump a piston indefinitely?

Wilhelm Schwake's 1926 pneumatic chamber tried to unload the labor onto the unwell occupant. The patent includes a drawing of a man standing in a box with his head protruding from the top and one hand on a handle on the front panel of the chamber. The entire panel served as a massive bellows that the patient, already winded from whatever ailment had landed him there, had to push and pull to inflate and deflate his compromised lungs. "We would not consider this respirator practical for severely ill patients, but the picture is a great favorite," wrote John H. Emerson, inventor of the most efficient and widely used negative-pressure ventilator in history, the Emerson "iron lung" (actually, stainless steel).

Two things ushered in the widespread use of negative-pressure ventilators: electricity and polio. In severe cases of paralytic polio, patients can no longer move their diaphragm and rib cage muscles and are thus unable to breathe on their own. Photos from the 1950s U.S. polio outbreak show iron lungs set up row upon row in "tank farms," some as long as football fields, to accommodate the crush of patients. I imagined a sort of human parking lot of misery. What could it have been like to spend months or even years in an iron lung inside a place like that?

Regina Woods contracted polio at the end of a high school summer vacation. It was a time when everything, she wrote in her memoir, seemed just about perfect. "I drifted along with the idea that I might someday be a Marine or a lady wrestler." She awoke one morning with an excruciating headache and pain in her neck and back. Soon she found she could no longer roll over.

By the time she got to the hospital, her breathing was labored. "Before I realized what was happening . . . I was in the huge cylinder with only my head sticking out."

I expected the next sentences to describe the unthinkable horror of having your body encased in a metal tank, spending your days and nights with a bedpan for company and no idea when or if you'd get out. I expected despair, horror, claustrophobia. Instead: "Such comfort! I was no longer struggling to breathe and the whole thing seemed simply wonderful. So wonderful that I cannot remember a night before, or since, of such sheer comfort."

Kathryn Black took a dimmer view. Her memoir about her mother, who contracted polio at age twenty-eight, holds no kind words for the "iron tomb" that encased her. Black describes a vacant, diminished human deprived of the touch of her husband and young children, lying awake at night listening to the "liquid noises" of the other patients and the "frantic clicking" they used to attract the nurses' attention—a sound "like that used in urging a horse to go forward."

For Arnold Beisser, a nationally ranked tennis player who contracted polio just out of medical school, the torment was existential. "I had become a 'thing,'" he recalls in a memoir. "Nothing was private." And: "Entry beneath my new metal skin was at the discretion of others. Those who attended to my body did things when and how they believed they should be done, and I seemed to have little or no part in this." Despite all this, Beisser comes around to a heightened appreciation of life's small pleasures and everyday wonders.

The Toomey Pavilion polio care facility in Cleveland, Ohio, published an in-house newsletter, *Toomeyville Jr. Gazette*. I read a few issues from the peak of the epidemic to try to get a sense of things there. The vibe was improbably sunny—helpful resources and warmhearted gossip. The dieticians were holding a concert. A nurse named Miss Brazos "likes strawberries with pickles." A staffer named Bob offered his services to anyone who needed a shave ("Legs a specialty"). Photographs showed iron lungs with aftermarket accessories to counteract the boredom—mouth sticks, reading periscopes, electric eye typewriters, mirrors mounted at an angle above the face, so patients could watch television and look around the room. Clothes and bedding and bedpans were changed via portholes in the sides of the tank. With a well-trained staff, these things could be done in minutes, in the manner of a pit crew swapping out tires. People seemed to have adjusted to their circumstances in a way I wouldn't have thought possible.

The July 1955 issue ran a short piece by a patient, Mary Ann Gasser, entitled "My Excursion to the Hospital Kitchen and Main Dining Room." "We started," she wrote, "by examining the bottle washing machine and also the pot and pan washer. . . . In a large mixer we saw potatoes being mashed in huge quantities. . . . I saw the inside of the oven and the doughnut and biscuit cutter." I'd been having one of those weeks where I felt pulled in four directions, bungling everything and falling behind. I found myself almost envying Mary Ann Gasser. I longed for the kind of unshattered focus that would allow me to marvel at the intricacies of a bottle-washing machine or succumb to the visual rhythms of an industrial potato masher. Perhaps the simplicity of confinement could be a kind of bliss, closer to a monastery than a prison. Maybe being in an iron lung is less horrible than it sounds. One way to know.

Mona Randolph contracted polio at twenty-one, in 1956, on the third day of a new job. Within three days, breathing had become a struggle and Mona was brought to a hospital and placed in an iron lung. For the next three months, day and night, she did not leave it. The March of Dimes helped pay for an Emerson unit to be placed in her home. She eventually recovered minimal use of one arm and enough diaphragm and rib movement to get through the day with occasional "sips" of air from a portable positive-pressure device. She returned to the iron lung at night to sleep.

Mona died in 2019. The iron lung stands where it always has, in a bedroom painted turquoise in the home Mona shared with her husband, Mark, on a quiet street in Kansas City. It is one of a handful of operational Emerson ventilators left in this country. I contacted Mark Randolph to ask if I might spend the night inside, letting it breathe for me. Mark said sure.

The Emerson's metal exterior is pale yellow, something I didn't anticipate, as most of the polio-era photographs are black-and-white. Mona called hers the Yellow Submarine. Inside is a narrow mattress on a cot that rolls in and out, to load the patient. If you've ever had an MRI, you might be reminded of that. Or you might just think that it looks like a hot-water heater. Mona's iron lung has a Wonder Woman sticker on the side, with the caption "No stunt doubles. This is all me." I'm pretty sure I would have liked Mona.

A few feet away from Mona's iron lung is a hospital bed where she would be transferred when she woke in the morning. A package addressed to Mark rests on the pillow. "That's Mona," he says, when he sees me looking at it. Mark is tall and solidly built, with a gentle demeanor and soft voice slightly out of keeping with his size. The disposition of Mona's cremains is one of the things on Mark's to-do list. I know this because the list is there on a closet door, in the form of yellow Post-it notes. "Car door repair." "Exercise 2× a week." "Make new friends." "Tool room clutter."

Mark and Mona were married for thirty-two years. He shows me photographs. They're a handsome pair. I hadn't been thinking about what Mark has been through until at one point I hear him say, "I went a week without a weeping spell, so I guess I'm doing okay." This is how this book thing goes sometimes. You think you're leaving to report a chapter on breathing machines, and then you arrive and there's a man standing beside an iron lung, saying, "I went a week without a weeping spell," and the man wants to tell you about the woman who once inhabited this iron lung you're so keen on trying, and so your plans scooch over and make some room.

I understand all the Post-it notes now. Mona was his passion, his pastime, and his to-do list. It's a big emptiness to fill.

I ask Mark to describe a typical day with Mona. It began by getting her out of the iron lung and over to the hospital bed. A pulley on a track on the ceiling helped, but mostly it was Mark. He recites the steps like an incantation, or a prayer, something repeated so often it comes out without thought. "Put her legs down, take the blankets off, give her arm a stretch. Take the straps for the lift. Sit her up and put the first strap under one arm. Then the other arm. Press the Up button, move her to the bed, lower her down. Stretch her legs out, reach behind you, grab the bedpan, put her legs up . . ."

There's another bed, a waterbed, in the room where Mark sleeps now. Once a week, per her doctor's allowance, Mona slept there too. "It was a way to let everything else go, to be close," he says. "To feel her next to me." And vice versa. Polio affects the motor nerves, but not the sensory nerves.

A couple of Mona's friends and former caretakers are coming for dinner and to help get me installed in the lung. Mark orders takeout. On the drive to pick it up, he tells the story of how he and Mona met. He was seated next to her at a dinner. Someone began saying grace, and the guests reached for each other's hands. Mona's paralyzed hand remained in her lap, so Mark reached over and picked it up. She told him later that most people wouldn't do that, that most are self-conscious around the very disabled. "All Mona wanted was to be treated like a normal person." He pulls into a parking spot and raises the car window. "So I passed the first test."

The caretakers arrive as we're setting the table. I mention that I'd been hoping to eat my meal while I'm inside the iron lung, as many polio patients had to do. Jane Buckley, who knew and cared for Mona for thirty-five years, informs me that that would be dangerous. You can easily inhale while food is going down, because the machine, not you, controls the timing of your inhalations. Bits of bacteria-laden food could wind up in my lungs, and that is a well-trodden path to pneumonia.

I assure her I'll be careful. Jane is a lovely person not easily swayed. Indomitable, you might say. Like a marine or a lady wrestler. There will be no eating in the tank.

After the plates are cleared, we head into Mona's bedroom. Jane slides the cot from the tank and arranges a blanket for me. One of those whimsical bird clocks warbles eight o'clock. Who can sleep at this hour? I climb aboard anyway. My head has to be maneuvered through the opening in the hatch at the front of the tank. If you were paralyzed, attendants on either side lifted and moved you toward the opening while another attendant guided your head through and tightened the collar. A 1944 booklet called "The Nursing Care of the Patient in the Respirator" includes a photograph of this. A nurse stands at the opening with one hand under the patient's head and the other over his face, as though birthing a full-grown adult. Because I can push myself along, the process is easier.

The good thing about the head remaining outside is that you don't feel claustrophobic. The bad thing is that your nose is going to itch, more fiercely than it has ever itched before, and your fingernails are trapped inside an iron lung. I ask whether there's an attachment for scratching your face.

"Yes," says Mark. "It's called a husband." Changing position or peeing in the middle of the night also required the husband. "Once or twice a night, I'd hear 'Maaa-ark! Maaaaa-ark!'"

For an iron lung to do its job, the seal around the neck can't be leaking air. Jane adjusts the collar clamp. It's padded, but I can feel it pressing on my neck.

"It's supposed to be snug," Jane says when I mention this. "Otherwise you won't get a full inflation of the lungs."

Snug! Har. "Is this how Mona had it?"

Jane slides two fingers alongside my neck. "No. Hers was tighter." She steps back. "Okay, we're going to lock you up now." She pulls a handle that seals the perimeter of the hatch. Mark flips a toggle, and the motor commences chugging. It occurs to me that I have never been able to sleep on my back and this is unlikely to be an exception.

I now understand why Jane nixed my iron lung picnic. You most decidedly are not in charge of your breathing. You will inhale when the machine makes you inhale, and ditto on the exhales. Should you try to defy the machine's rhythm, there will be snorting and mild panic until you fall into line. You cannot talk on the inhales. This is true even outside of an iron lung (go ahead, try it), but normally you are able to pause your inhale to speak or laugh. I keep trying to do this, and each time, the machine cuts me off mid-syllable, the silence preceded by a strangled *cackh-guh!*

I'm sucking in big, long steady breaths. "I think it may be set too h— *cackh-guh*." Indeed, it is set as high as an Emerson iron lung can be set, because that's what Mona needed. Recall the scene in *The Princess Bride,* Westley trapped on his back with The Machine cranked to its maximum. Mark reaches down to adjust a setting.

Better. Now I could almost imagine sleeping in here. Except for the collar. It's unusual to feel yourself breathing deeply, luxuriantly, and at the same time feel like you're being strangled. I note that the collar makes a lurching flutter each time the iron lung pulls in air. "That's because it's not tight enough," Jane says. She adjusts it. The two of them walk to the bedroom door.

"It's very restful," says Mark. "You'll be asleep in a minute."

Jane flips off the light. "Good night!"

"—*Cackh-guh*."

I lie in the dark, listening to the very deep and deliberate breathing that is, weirdly, coming out of me. I raise my knees. I flatten my knees. The blanket slides off. Some minutes pass. Not a lot of them. The clock strikes nuthatch.

"Maaa-ark!"

Mark sticks his head through the door. In short utterances between Darth Vader inhalations, I relay that I've probably had enough. He walks over and flips the motor switch.

"How long was I in?"

Jane opens the hatch. "I'd say five minutes."

"That's it?"

"It may have been eight or nine."

We go back out and sit around the dining room table. I'm curious to know how Mona coped. Obviously it's a different experience for someone who otherwise can't take in enough air and spends the day working hard to do so. Breathing, for Mona, was exhausting, and one sleeps better exhausted. Jane sums it up this way: "She was so relieved to get in it, and so relieved to get out of it in the morning."

The alternative, for someone like Mona, would be some kind of positive-pressure device—air pushed down the windpipe. A medical leaf blower. Some people use a tube that feeds through a hole cut in their windpipe via the front of the neck. Others use a mask over the nose and mouth, similar to a CPAP (continuous positive airway pressure) machine—a device so unbeloved that, according to a 2008 study, 46 percent of sleep apnea patients who've been prescribed one either abandon it or never once put it on. Mona was able to get by during the day with sporadic shots of air from her "sip tube." That's as far as she was willing to go with positive pressure.

"Mona *hated* positive pressure," Jane says. There is plenty to hate.

With the success of the 1950s polio vaccination campaigns, iron lungs trundled into obscurity. Soon after, positive-pressure ventilators arrived on the scene. During the Korean War, medics had discovered that pushing massive amounts of fluids intravenously would keep a hemorrhaging soldier's blood pressure up so he wouldn't go into shock. The technique saved lives but resulted in a fluid overload in the lungs, called "wet lung syndrome." This was dealt with by using a machine with sufficient pressure to push air through the fluid. Positive-pressure ventilators, as they became known, were widely used during the Vietnam War, and soon civilian hospitals began using them too.

The story was relayed to me by Norma Braun, who lived through that era. Braun, in her eighties now, is a clinical professor of medicine and critical care at Icahn School of Medicine at Mount Sinai, in New York. She specializes in pulmonary care for people with disabilities that compromise their breathing—including, for many years, Mona Randolph. Braun was an intern with the Bellevue Hospital Chest Service during the polio era and remains a fervent advocate of negative-pressure ventilation. While no one, Braun included, wants to resuscitate the iron lung, there are newer, much abbreviated negative-pressure devices in development. A device called the Exovent, under development in the UK, encloses only the chest, leaving the arms and legs free.

I visited Braun in New York before heading to Kansas City. What, I wanted to know, is the problem with the positive-pressure ventilators of today's critical-care units? Braun took a deep breath and exhaled slowly. (When you're writing about respiration, you notice things like this.) As if to say: where even to begin.

Intubation. Let's begin there. Because the tube passes between the vocal cords, patients can't talk. They can't communicate easily with their family or tell a nurse when they're in pain. Nor can they swallow, so they have to be tube-fed. More gravely, intubation carries a risk of pneumonia. "When you push a tube down a patient's throat," Braun said, "you are injecting into the lungs the organisms that are in the mouth." ICU patients likely have a heavier than normal load; flossing isn't a priority when you're playing dodgeball with death. After ten days, one study found, 90 percent of patients earned a dental plaque score of 3 (the worst), even though 70 percent had gone in with a score of zero or 1 (the best). Ventilator-associated pneumonia is the most common hospital-acquired infection in ICUs. The mortality rate is as high as 50 percent.

Then there's the inflating. Positive-pressure ventilators don't expand the lungs in the same manner as natural breathing—the original negative-pressure ventilation. The difference was demonstrated in a study comparing negative versus positive pressure on a lung hooked up to a device engineered in the BREATHE Center at the University of California, Riverside, School of Medicine. "Not only is positive pressure quantifiably different, it's astonishingly different," said Mona Eskandari, who helped design the system and run the study. The upper portion of the lungs overinflates—picture blowing up a party balloon—and the lower portions barely inflate at all. So you have some of the lung being repeatedly overstretched and pummeled by hard-hitting air, while other parts lie fallow—to the point of partial collapse. With the top of the lungs doing all the work—a condition known as "baby lung"—blood oxygenation drops, and mucus builds up in the collapsed regions below. Mucus being, as Braun put it, "a banquet for bacteria." Yet more risk of pneumonia. The antibiotics used to treat pneumonia can create their own problems. Aminoglycosides, a class of potent antibiotic used on patients with life-threatening infections, cause some degree of hearing loss in 10 to 66 percent of patients, depending on which drug is used.

Hard-hit COVID-19 patients had it especially rough. The virus caused the lungs to stiffen, making them harder to expand and less able to absorb oxygen. But if you raised the ventilator pressure to the level needed, you risked further damaging them. "When you push air into a very stiff lung you burst the alveoli, and that's when you die," Braun said. Alveoli are the tiny sacs where the life-sustaining exchange of carbon dioxide and oxygen takes place. To avoid rupture, the breathing rate is speeded up: same amount of oxygen, delivered via more frequent, less forceful breaths. While this is safer, people don't breathe comfortably that way. Patients panic. They try to take in more air, and they may end up breathing against the ventilator. "That is one of the most damaging things you can do to a lung," says C. Lee Cohen, a critical-care medicine instructor at Brigham and Women's Hospital, whose COVID-19 care protocols she developed.

To avoid this "dyssynchronous breathing," ICU patients on mechanical ventilation are often sedated. Long-term sedation causes delirium that may last weeks or months, putting the patient at risk of falling or choking. And if you can't get patients sedated sufficiently to keep them from trying to breathe on their own, you may need to take over the controls by knocking the diaphragm and rib muscles out of commission—that is, paralyzing them. Cohen works hard to avoid this. Lying paralyzed for days quickly and dramatically weakens muscles—including, of course, the diaphragm—to the extent that it will be harder to get off the ventilator and breathe on one's own, harder even to get up and about.

"But what's the bigger risk?" Cohen said when we spoke by phone. "That they're not going to walk again? Or that they're going to die within twenty-four hours. That's the call you make."

I asked Cohen about negative-pressure devices like Exovent. She didn't foresee a return to negative-pressure ventilation in ICUs. It wouldn't suffice, she said, for patients with severely stiffened lungs or narrowed airways. Those lungs are harder to inflate, and the damage prevents them from absorbing enough oxygen. (Negative-pressure ventilators don't deliver surplus oxygen; the patient is breathing room air.)

Relying on an Exovent-type device in an ICU setting would present logistical problems. How do you use a defibrillator or a bedside X-ray machine on a chest enclosed in a box? The Exovent has access ports, but as the company's CEO, Ian Joesbury, conceded, it probably isn't suitable for someone who needs complete breathing support or access to the chest for taking X-rays. And the models I've seen don't monitor levels of oxygen and carbon dioxide in patients' blood—a critical component of ICU care.

Both Cohen and Joesbury saw potential in some combination of negative and positive pressure. The negative pressure could address the "baby lung" problem, by encouraging a more natural distribution of air throughout the lung. And that could allow staff to reduce the pressure to a safer level—without speeding up breathing and requiring sedation. Supplemental oxygen could then be delivered by a nasal tube or a mask—as with the CPAP devices used by people with sleep apnea.

For people like Mona Randolph, with chronic breathing difficulties that arise not from damaged lung tissue but damaged nerves, negative pressure can be a godsend. Norma Braun had a prototype of a simple wearable negative-pressure ventilator on hand for me to try when I met with her in New York. Called the Venti, it consists of an abbreviated airtight bodysuit with a flexible shell that fits over the chest, plus, of course, a motor-powered pump. It was gentler than the iron lung, an assist rather than a *coup*.

The simplicity of negative-pressure ventilators can be an advantage. They're inexpensive to manufacture and easy for medical staff to learn to use. Simplicity and affordability mean better accessibility—for home users with chronic breathing issues as well as for patients in developing nations and clinics that may lack the funds for a positive-pressure system. It's a truth that's clear-cut in much of the world: if the cost of high-tech medical care makes it inaccessible to all but those who can afford to travel elsewhere, then low-tech care isn't inferior at all.

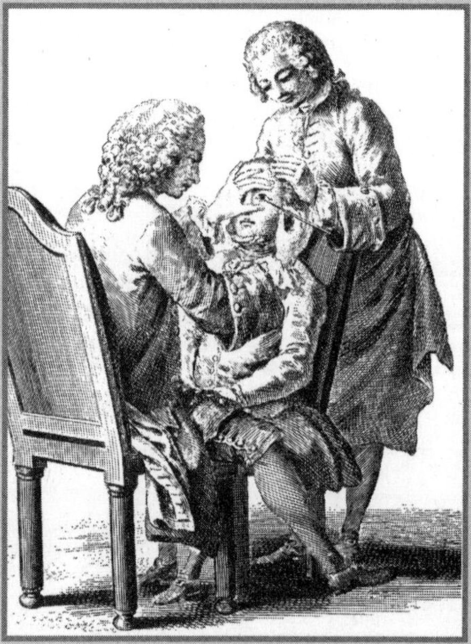

11
The Mongolian Eyeball

With Cataract Surgery, Sometimes Simpler Is Better

The Mongolian countryside is government land. Instead of the patchwork of farms and fences that divide an American countryside, the land is an unbroken expanse, more a blanket than a quilt. On this rolling and largely treeless nation, Mongolian nomads—who make up around a fifth of the population—are free to set up camp and graze their herds where they wish. Likewise, you may drive wherever you wish. The countryside has few roads. Or is one big road. Feel like visiting the folks who live off by that stream in the distance? Turn the steering wheel and aim the car.

That is how I came to be standing at the door of a ger in a grassy expanse hundreds of miles from a city. (Mongolians say *ger* rather than the Russian *yurt*.) I'm traveling with a group of ophthalmologists, Mongolian and American, from the eye health nonprofit Orbis International. We are driving—or were until a moment ago—to Khuvsgul, one of seven rural provinces where Orbis has set up and equipped surgical training programs and facilities.

I had expressed an interest in nomadic life, so the Mongolian ophthalmologist in the driver's seat, Uno Bayaraa, chose a ger at random and went left, off the roadway. Such is nomad hospitality that you can knock on the door of a stranger's ger and explain that your companions would like to check it out. "It's not awkward," Uno insisted. (It was a little awkward.) For simplicity's sake, he introduced us all to the man in the doorway as eye surgeons. As though this were normal, as though ophthalmologists roamed the countryside in herds, like yaks, the man invited us in.

When I first heard about Orbis, I assumed we'd be flying, not driving. Orbis runs the Flying Eye Hospital. I imagined nomads on horseback lined up on a tarmac, waiting to have their eyes repaired. It's not quite like that. The Flying Eye Hospital is a teaching hospital, with a classroom and a surgery simulation lab in addition to its operating room. Flying surgeons to remote areas to operate for a week or two is impactful for relatively uncommon afflictions like cleft palate; less so with something as ubiquitous as cataract. (If you live long enough, you will have cataracts.) Far better to train and equip local eye doctors to do the procedures. "If you can teach a man to fish . . ." That's what the head of communications said when I first contacted Orbis. *Who is Amanda Fish?* I thought, and then I remembered the reference: *Give a man a fish, and you feed him for one day. Teach a man to fish . . .*

Orbis has been teaching Amanda Fish for forty-two years, in nearly every country and territory in the world. This week, the recipient of Orbis largesse is an eye surgeon in the city of Murun, our destination in Khuvsgul. She'll be observing and assisting with a series of cataract surgeries being undertaken by another Orbis-trained surgeon, who is traveling with us today in a second car. The Orbis mission is to prevent avoidable blindness and vision loss, and cataract is the leading cause of it.

We duck through the doorway of the ger and sit down in a circle. The man's daughter gets up to make tea. It's a crowd: all of us, the man, his daughter and wife, her mother, and the majority holdings of a butchered sheep. As we entered, the man slid the carcass back to make room but did not drag it outside. Most of the things inside this ger are livestock-related: A Best Herder award from the provincial government. A bowl of fresh sheep's milk curds. A saddle and a hanging assortment of bridles, lots of bridles. They have bridles the way American businessmen have ties.

Animals are a nomad's livelihood: meat and dairy, to eat and to sell, plus, in Mongolia, goat hair for cashmere. Sheep are a kind of currency. "You sell ten sheep to buy a motorcycle," Uno said earlier, in the car. I asked him what the exchange rate is. I was joking, but he gave a straight answer. "Right now, one sheep equals eighty to one hundred dollars." The prison sentence for stealing a sheep, he added, is as much as ten years.

As we get ready to leave, the man steps outside with a pair of binoculars. I follow his gaze to a far hillside with a barely visible stubble of sheep and goats. Binoculars: a better set of lenses. I wonder about his own lenses. Cumulative exposure to the sun's ultraviolet rays is a risk factor for cataracts. Nomadic Mongolians spend long hours outside under a vast CinemaScope sky, a sky so combatively bright that you want a visor even when the sun is behind you. Shade is as rare as salad. In almost 40 percent of cases here, the cause of blindness is cataracts.

Modern cataract surgery puts a better set of lenses directly into your eyes. In the United States, the surgery is so dependably safe and effective, the recovery time so short and the side effects so minimal, that twentysomething myopes are having their own perfectly clear but misfocusing lenses exchanged for prescription intraocular lenses (IOLs) that will render them 20/20. They're opting to have cataract surgery without the cataracts.

The rural regions of the developing world are a long way from all that. This past week in Ulaanbaatar, the country's capital, the patients I met, all of them from outlying towns, had let the condition progress to the point where they could barely see through the lens. The dearth of well-trained eye surgeons in the countryside is to blame. It's not just the lack of care but the quality of it. When someone's cataract surgery goes poorly, word gets around. Yesterday at First Central Hospital I spoke with a man about to have surgery to remove a late-stage cataract. He related that he had not worried about his heart surgery but had for years resisted cataract surgery.

For seventeen centuries, it was well worth resisting.

The ophthalmological term "cataract" refers to both the disease and the clouded lens itself. It derives from the word's original meaning. A cataract is a large waterfall, and the water, as it thunders down, turns from clear to opaque. (In Mongolia, perhaps because there are few waterfalls of note, people don't say "cataract." They call it "cloudy lens disease.") The original surgery for cataract has a similarly descriptive name—"couching," from the French *se coucher,* to lie down. The cataract—that is, the opacified lens—wasn't removed from the eye, as it is today, but pushed to the bottom of the eyeball. It's a ludicrously genteel term for it—as though the lens were gently guided to a divan like a fainting demoiselle.

Hardly. Here is Celsus, Roman chronicler of medicine, describing couching in *De Medicina*: "The needle, sharp-pointed, but by no means too slender . . . must be thrust in." The moorings that hold the lens in place are cut, and then comes the couching: "By degrees work the cataract downward below the pupil." Don't be shy! "It must be pressed down with a considerable force." And held there for what must have seemed to the patient like several lifetimes: "For as long as it takes to say four or five Paternosters," advised one twelfth-century medical manual.

The immediate effect of couching was impressive, at least to onlookers. With the opaque lens moved aside, suddenly light could again reach the retina, and vision, albeit out of focus, was restored. (Keep in mind, the lens supplies only a third of the eye's focusing abilities. Two-thirds derives from the cornea. Laser vision correction reshapes the cornea but does nothing to the lens.) Couching was equal parts theater and surgery. The coucher milked the unveiling. "The operator tested his patient's vision immediately after the operation by holding up fingers, coloured cloths . . . or other common objects, for triumphant identification," wrote Robert Henry Elliot, author of a slim 1918 volume entitled *The Indian Operation of Couching for Cataract*.

A day or so later began the trouble. Pain, inflammation, infection. The eye's lens resides in a protective capsule. Nothing gets in, including elements of the person's own immune system. If the coucher's knife breached the capsule, the immune system would encounter the lens for the first time and assume it to be an intruder. One reason cataract surgery was slow to incorporate replacement lenses—IOLs weren't routinely used in cataract surgery until the 1970s—is that it took a long time to find a material that wouldn't provoke an immune assault. (It was a chance discovery. During World War II, Royal Air Force eye surgeon Harold Ridley noticed that fragments of shattered Plexiglas jet canopies that lodged in airmen's eyes were well tolerated, and that it was safer to leave them there than to try to dig them out.)

Around the mid-1700s, couching—or "depression," as it was later called—gave way to extraction. Surgeons would cut a large semicircular incision through which the lens and capsule would be pulled out using forceps or, later, a freezing-cold probe that would fuse to the lens like a tongue on a metal pole in winter. Up through the 1960s, this is how cataracts were removed. The surgery required general anesthesia, multiple stitches to close the incision, and an eight-to-ten-day hospital stay, during which time patients were confined to bed, sometimes with sandbags on either side of their head to keep it still.

Regardless of whether a lens was pushed down or pulled out, patients were left uncomfortably farsighted. You don't need a lens to see, but you need it to see well. Until IOLs were perfected, people left lens-less by cataract surgery had to wear glasses with thick aphakic "Coke bottle" lenses.

"Remember the comedian George Burns? He had that surgery." This from Orbis's VP of clinical services and technologies,

Hunter Cherwek, the man crammed beside me in the back of the car. Cataract surgery explains Burns's trademark eyewear. (For a more contemporary example of aphakic lenses, picture Stephen Root with his red stapler in *Office Space*.)

We are off-road again, on an unmarked detour. The road to Khuvsgul is under construction and seems to have been so for some time. The tire tracks of the original detour are rutted out, with a skein of newer tracks on either side. Because the scenery is so spare, the things one encounters stand out. They're like actors on a bare stage. A bleached skull. Three horses on a ridge. A tree coming out of a rock outcropping like a fascinator on a coronation-day hat.

As recently as thirty years ago, Cherwek says while we jounce across a stream, couching was practiced in rural parts of Mongolia's neighbor to the south, China. In 1999, a team of researchers surveyed the main causes of blindness in Doumen, a county in rural southern China. Couching ranked third.

"Chairman Mao was couched," Cherwek says.

Since we left Ulaanbaatar, Cherwek has been a continuing source, a thundering cataract, of eyeball lore. It's all through my notes. "Rhinos are myopic."* "Judi Dench echolocates." "The lens is the only part of the body that doesn't get cancer." Orbis staff have a term they like to use: Cherwekipedia.

---

* The internet will tell you that the rhinoceros is extremely nearsighted, so much so that it occasionally charges termite mounds, mistaking them for rival rhinos. However! Sometime in the early 1990s, a team of Cornell University biologists examined four rhinos—three captive white rhinos and one wild black rhino—and found none to be nearsighted. The rhinoceros, they determined, is in fact very slightly *farsighted*. How do you give an eye exam to a wild rhinoceros? You shine a spotlight into its eyes from the safety of your Kenya hotel room balcony, use your binoculars to note the maximum distance at which eyeshine can be observed, and apply a variant of the point spread retinoscopic technique, duh. The study, entitled "Refractive State of the Rhinoceros," was funded in part by a grant from the National Institutes of Health, whose reviewers are sufficiently farsighted to see value in such things.

This is not to suggest that Hunter Cherwek is a know-it-all. Not at all. He's a know-a-lot, and he likes to be helpful. He may be the most helpful human I've known. If you drop something, he will pick it up before you notice you've dropped it. He has insisted on taking the middle seat for the entire twelve-hour drive to Khuvsgul. He keeps his knees together so as not to infringe on others' legroom, and he manages to hold them that way while napping. He's considerate *even in his sleep*. Cherwek is a man born for charitable pursuits, a fact he's known since college. The day he finished his ophthalmology residency, he joined Orbis. The world and its eyeballs are the better.

Horribly, couching carries on in parts of rural Africa. Orbis hopes to eradicate procedures like this. It is one thing to eradicate a disease, but how do you stamp out a surgical procedure? Orbis's approach is to go into these areas and train doctors to perform better, safer procedures. "Quality drives demand," Cherwek likes to say. Along with training, Orbis donates equipment: surgical microscopes, diagnostic equipment, boxes of IOLs.

Cherwek leans over to see out my window. "Are those wild camels?" The conversation goes sideways as we try to remember the name for a double-humped camel. Bavarian? Bacitracin? Bactrian!

The operating table in the Murun General Hospital is made of wood. It has no wheels and no padding for patients to lie on. It's a table one might find in a rustic cottage, with people sitting around it having a meal. Behind it, against one wall, is a chest of drawers, also not something you would find in a medical supply catalogue. The patterned laminate on one of the drawers has English words in cursive. I walk over to look more closely: "Sweet love."

As basic as this OR is, it holds more than what is actually needed to restore a cataract patient's eyesight. A manual cataract operation can be done with six handheld instruments. Cherwek points out that the only thing plugged into a wall right now is the microscope that the surgeon looks through while operating. Even that isn't strictly necessary. Practitioners can use loupes—glasses-mounted magnifiers—and a headlamp.

Manual small-incision cataract surgery—the procedure I'm watching this morning—is not something you'd see in a state-of-the-art ophthalmological surgery suite. There you would see "phaco": phacoemulsification, pioneered by the late Charles Kelman. Kelman was an ophthalmological inventor who took his inspiration from dentistry. Like a hygienist's plaque buster, a phaco wand emits ultrasound. It vibrates the lens into a chowder of small pieces that can then be suctioned out through an incision small enough to require no stitches— about two millimeters long. Earlier dental inspiration was less wonderful. "I theorized that if a dentist could drill a tooth, I could drill a cataract," writes Kelman in *Through My Eyes,* a title that sounds more perforative than personal once you've read a few chapters. It went poorly. "I watched in horror . . . as the cat's iris . . . wound itself around the rotating drill." Other failed attempts employed an optical Home Depot of "manual disintegrators": "high-speed cutting needles, a miniature blender, drills, tiny meat grinders . . ." Sweet love!

A phaco machine is expensive, and surgeons and residents must be trained to use it. The operating room must be reconfigured. You have to upset the status quo. At the moment, phaco is available only in Ulaanbaatar. Elsewhere, manual small-incision surgery is the solution. "It's low-cost and easy to teach," said Orbis volunteer faculty and cataract demigod David Chang, whose Los Altos, California, practice sits at the other end of the spectrum. "There are fewer complications and they're easier to deal with." That's especially important in places like Mongolia, where people tend to let their cataracts progress to an advanced state. A very old, very hard cataract is more difficult to completely and safely remove.

It is a matter of greatest good. It may take decades to get Mongolia's rural provinces set up for phaco. And in that time—by training and equipping local eye doctors to remove cataracts manually—you can restore vision, good if not perfect, in thousands of patients on the downhill slide to blindness.

The woman on the operating table has traveled from up near the Russian border, about a hundred miles north. She is draped with a blue sterile cloth that has a clear plastic window over the eye. The surgeon operates through a slit in the plastic that is held open, along with the eyelids, by a small speculum. The way the plastic is pressed to the patient's brow makes me think of those slipcovers people's grandmothers used to use to protect the good furniture.

Performing the surgery is Enkhzul Damdin, the Orbis-trained ophthalmologist who has traveled with us from Ulaanbaatar. The second surgeon, the trainee, is observing through a second eyepiece, a sort of microscope sidecar. Both are women. Ninety percent of Mongolia's eye surgeons are women, not because the nation is a powerhouse of women's liberation but because the pay is so low. "There is an impression that it's easier than manual labor," says Chimgee Chuluunkhuu, an ophthalmologist who is also the director of Orbis Mongolia (and my translator for today). "Truck drivers are paid three times as much as eye surgeons," she adds. Chimgee is stylish to an extent I couldn't suitably appreciate when we'd spoken on Zoom, because the best of it happens below her ankles. Today: chunky-soled metallic copper sneakers. Yesterday: sequined Vans in steel blue. I ask her how she decided to become a doctor. "I liked chemistry," she says, then reconsiders. "Actually, not really *liked*. I was good in chemistry."

Damdin is getting ready to undertake the trickiest part of the operation: capsulorhexis. Using forceps, the surgeon peels away the central part of the top of the capsule, creating the internal opening through which the lens exchange will take place. The capsule is critically important because it holds the lens in place. It's also insanely delicate. It's as thin as the skin of a cherry tomato, which surgeons sometimes practice on. Using tweezers, try to tear away a neat circle of tomato skin. Nine of ten early tries, the tear will take off down the side of the orb. If that happens during cataract surgery, the new lens will be unstable. It may slip out of position. The high-tech version of a cherry tomato is a virtual reality surgery simulator recently co-created by Orbis and now up and running in their Ulaanbaatar training lab. With it, surgeons can practice hundreds of times before attempting the maneuver on a surgical simulation eye, each of which costs about $30.

As basic as it is, manual small-incision cataract surgery is a significant improvement on the state of the surgery in past years here. Older techniques were done through a larger incision that required stitches to close it. Patients would have to come back to have the stitches removed, and some would put it off, and the stitches would get infected. Or they'd cause dimpling on the surface of the eye, which left the patient with astigmatism.

Complications like these were once common in the U.S. as well. That's why ophthalmologists used to advise patients to hold off until their cataract was "ripe." "With the large-incision surgery," Cherwek told me, "they couldn't get a good percentage of their patients to 20/40 or 20/30, so they'd wait for the cataract to progress." Only when a patient's vision had declined to, say, 20/80 could the surgeon be relatively confident that the operation would actually improve their vision.

Chimgee alerts me that the lens is about to come out. It's a little like the final moments of a birth: the anticipation and the rapid home-stretch glide. The first time I watched it happen, in the OR yesterday in Ulaanbaatar, I got a little swept away.

There it was, abruptly, the lens of a human eye, a miracle of physiology, just sitting there where the surgeon had parked it, on a square of gauze on the instrument tray. Without really thinking, I folded over the gauze and slipped it in my pocket to examine more closely later. Cherwek calmly pointed out that I would be taking human tissue without consent. Which could cause the hospital to lose its license. And was, should the cataract leave the country, a violation of human trafficking laws.

"Okay, okay." I handed it over.

The Orbis photographer, Geoff, was enjoying the exchange. "My wife went to Mongolia, and all I got was this cataract."

On a reality TV show, this would be the reveal: the moment the bandages come off. The Orbis communications staff is poised to capture it, cameras and iPhone voice recorders in hand. These scenes are valuable for fundraising—a newly sighted patient hugging his surgeon or weeping with joy as he sees his grandchild's face clearly for the first time.

The eyeball-under-a-slipcover, from yesterday, turns out to be a herder with eight children and "a complete set" of livestock (sheep, goats, horses, cows, camels). Her face is tanned and deeply lined. It's hard to know how much is from sun and how much from age. I love her look. Her pants are tucked inside tall black leather work boots. Her sweater is primary-color green and her earrings speak to the silver strands in her black hair. She's sitting on a chair in the middle of the room. People walk in front of her, unaware that she is trying to read an eye chart. Meanwhile I'm trying to read her. I had imagined some heightened buoyancy to her mood.

The herder leans forward in her chair. "It might be a three?" Chimgee still translating.

It is a three. It's been one day and already she's reading the 20/30 line! I'm more excited than she appears to be.

"People from the countryside are very reserved," Chimgee explains. "They don't show their emotion."

Nomadic Mongolians live hard lives. Reining in anger or frustration is a way to cope. The good feelings get submerged along with the bad, I suppose. There was a moment in Ulaanbaatar yesterday that summed things up for me. Geoff was photographing a woman's reaction to the clear vision she was suddenly enjoying. "How about a smile," he said. And she said, "I have no teeth."

The herder in the green sweater gets up from the exam chair. She is bowlegged, and not in the rounded, comical manner of old movie cowboys. Her knees jut out to the side, bent at sharp angles. Her legs are math: Less Than, Greater Than. Every step must be painful, though her expression doesn't reflect this.

"People from the countryside never complain," Chimgee says. "The women here are very tough." She tells a story her advisor shared years back. Before there were sterile surgical drapes, patients' hair would be wrapped under a cloth that was clipped together to secure it. "This one lady, they clipped her skin too. And she said nothing. She just thought, *It's surgery, I guess it's supposed to hurt.*"

At our prompting, the herder shares her story. Chimgee summarizes. She was widowed at thirty-three, when the youngest of her children was eight years old. She climbed hillsides after goats. Bathed in the river, peed outdoors even when the temperature was minus 50. Her days would begin at 5:00 a.m. "Milking, milking, milking. Liters and liters. Then you need to boil it, otherwise it will spoil. Then you have to do all this dried yogurt stuff and make cheese." I ask how she spent her weekends. She laughs at this. "No weekends!" Now she takes care of her grandchildren. Nomadic families send young children to school in the town centers during the summers. She goes along to care for them. Someone asks how many grandchildren she has. Chimgee listens and reports back. "She doesn't know exactly."

Without her cataract surgeries, the woman in the green sweater would be nearly blind. Blindness is the end point of cataract and an untenable situation for a Mongolian nomad. Herder families are self-sufficient. Survival depends on everyone's hard work. A family member who can't help with chores becomes one themself. You're either easing the burden or making it bigger.

The traffic on the road back to Ulaanbaatar is slight. There are more horses in view than cars. We pass a young man texting while riding. The smells of the land blow in through the open windows, sage and goat. Cherwek once again in the middle seat, a captive to my list of questions.

One says simply: *Presbyopia?* Cherwek explains that the middle-age descent into reading glasses is caused by the lens growing more and more rigid over time. This happens because the body keeps adding layers to it; by middle age, the lens is crammed so tightly in its capsule that it's too stiff for the eye muscles to manipulate into the various thicknesses that supply clear vision at different distances. "The lens," reports Cherwekipedia, "is the only part of the human body that continues to grow throughout life."

Cherwek pulls his hat down over his face to sleep. We drive in silence for a few miles.

"Hey, so why does the body keep adding layers to the lens?" It's the second time I've interrupted his nap. The hat stays down.

"I'll ask God."

A single-focus intraocular lens, like the one the herder received, does nothing for presbyopia. For that you need either a multifocal lens—with two or three zones of correction, as progressive glasses lenses have—or, ideally, an accommodating lens. An accommodating intraocular lens would be a lens that somehow delivers a range of focus as complete and nuanced as what we had and took for granted as teenagers. It's the Holy Grail of cataract treatment, and it remains—despite some companies' claims—unattained.

Understandably. Accommodation is no simple feat. The ring-shaped ciliary muscle is attached to the lens by spoke-like fibers (called zonules). When the ciliary muscle is at rest, the spokes are pulled taut, which in turn pulls the lens into a slightly flattened shape. When you need to focus on something farther away, the ciliary muscle contracts, allowing the spokes to slacken and the lens to thicken—thereby becoming more powerful. For complicated reasons, there's no way to re-create the gorgeous interplay of lens, muscle, and zonules. Instead, companies have tried to design lenses that would change thickness via other mechanisms. "There's a whole graveyard of lenses that tried to mimic natural accommodation," says Malik Kahook, chair of the Orbis medical advisory committee, whom I spoke to before coming to Mongolia. "Five or ten companies that came and went."

Julie Schallhorn, a surgeon and professor of ophthalmology at the University of California, San Francisco, and the lead author on an American Academy of Ophthalmology report on accommodating lenses, walked me through the graveyard. There was the one with the four fluid-filled "pillows" that squeeze their contents into the center of the lens when the ciliary muscle contracts, thereby making the lens thicker. (The ciliary muscle retains some movement after cataract surgery.) "Used to be available in Europe," she said. "Who knows what's happening with it now." Another approach was to use a flexible gel in place of an intraocular lens. "There are toxicity questions," she said of that one. Schallhorn was speaking to me from her home, because the au pair was out with appendicitis. Children cycloned through the room.

Something called the Synchrony had two lenses that moved relative to each other. "That totally didn't end up working. Walter, share with your sister." Walter, age seven, was battling his sibling over a stash of Valentine's Day candy. I could hear them in the background.

Schallhorn got back to it. "Then there was a company in Japan that tried putting microelectronics on their lens, with, like, a liquid crystal to change the power of the lens. To my knowledge—"

"Walter's being a stinker!"

"—to my knowledge it has not gone into any human eyes."

"I am *not* a stinker."

Even the far simpler multifocal lenses have fallen short of expectations. Halos around lights and other minor visual aberrations are still common. In exchange for being able to read both books and highway signage, you lose contrast sensitivity—meaning it's harder to discriminate between contrasting colors. Schallhorn compared it to turning off a lamp and struggling to read in the semidarkness. She recited a quote from Jack Holladay, a name you and I would know if we were eye surgeons. "'In optics, there's no free lunch.' If you want to blow bubbles, go to the backyard."

The majority of cataract patients still opt for simple, dependable, Medicare-covered single-focus intraocular lenses—the kind being installed in the operating rooms of rural Mongolia. They see perfectly at one distance and when they want to see clearly at the others, they put on glasses, the tried-and-truest replacement lenses of all.

Before I came to Mongolia, I thought that the reason we have trouble reading when we hit middle age is that our focusing muscle, the ciliary muscle, becomes too weak to do its job. I now know that's wrong—it's the lens itself that changes, not the muscle.

That's good, because there will likely never be an artificial ciliary muscle. I once asked Rick Redett, a surgeon I met while researching a previous book, what he thought was the trickiest body part to replace. "Things that contract" was his answer. Muscles. In particular, the ring-shaped muscles. He mentioned the ciliary muscle and the orbicularis oris, which surrounds the mouth. Surgeons have tried to replace the latter by transplanting muscle from the thigh or the back. "The results," he said, "were never remotely close to what the natural muscle does."

He also mentioned the anal sphincter. When things go wrong in that department, an altogether different strategy is called for.

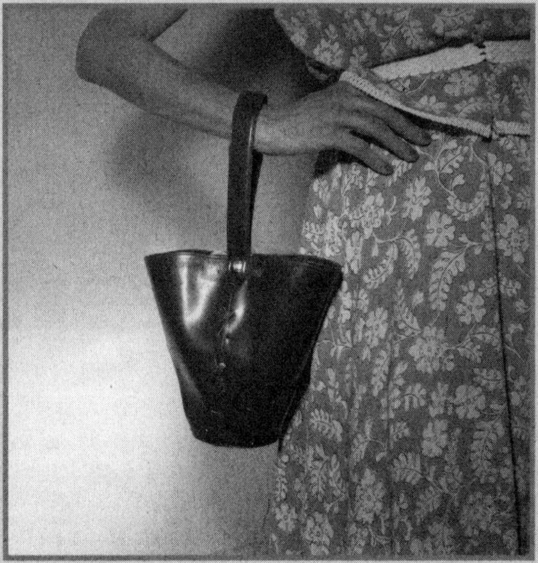

12

The Last Six Inches

Battling the Stigma of Ostomy

*S*toma, plural *stomata*, comes from the Greek for "mouth." More broadly, the word is used for biological openings: holes for eating, holes for breathing, holes for excreting. Wherever an organism opens its inside to the outside world. In the gastroenterological universe, *stoma* most often refers to a surgically created opening through which waste is diverted—into an external pouch, typically: an artificial rectum. If a stoma is opened along the colon, the surgery is called a colostomy. Along the small intestine, it's an ileostomy. (*Ostomy* is the umbrella term.)

Stigma, plural *stigmata,* also comes from the Greek. Originally it meant a physical mark, but the word has come to refer more abstractly to a mark of shame. There may be no more potent medical stigma than the one that marks ostomy. United Ostomy Associations of America (UOAA) seeks to change this. They would like to—borrowing their phrasing—flush the stigma.

To normalize a closeted condition takes time and talk—open talk, public talk, talk without euphemism. There are many ways to do it, and UOAA has tried most. A National Ostomy Awareness Day was proclaimed in hopes of attracting media coverage. My Google Alert for "National Ostomy Awareness Day" in 2022 yielded a single media mention, in a small-town paper in central California.

"They won't cover it," says Ed Pfueller, the UOAA communications and outreach director. "They think it sounds gross."*

Pfueller would love to recruit a celebrity, someone relatively cool, like Pearl Jam guitarist Mike McCready, who has done fundraisers for the Crohn's & Colitis Foundation. He knows of some people, and the organization has discreetly approached them. "They said not to tell," he told me when we first spoke, by phone. So Pfueller makes do with the dead: Bob Hope, Fred Astaire, former President Eisenhower. The late Matthew Perry, in his 2022 memoir, supplied some celebrity un-endorsement, declaring his nine months with a colostomy bag "hellish."

Each year on National Ostomy Awareness Day, a few of the state-level ostomy associations hold a 5K fundraising run. People without stomas are invited to participate and to wear a pouch while doing so, to boost awareness and empathy. I had planned to go to the largest one, in Durham, but Hurricane Ian was on track to hit North Carolina on Ostomy Awareness Day. With two days to go, the run was canceled. Ed Pfueller can't catch a break.

Pfueller recommended the Arizona run instead. He said it should be almost as well attended, with 194 ostomates and supporters signed up. (The term *ostomate* is used widely within the community. Like *roommate* or *shipmate*, it conveys kinship and a small shared universe.) "The Giant Inflatable Colon will be on-site," he added. If you know me, you know that that is, for better or for stupid, the sort of thing that tips the balance. I booked a flight.

---

* Even media people at ostomy supply companies seem to shy away. Four phone messages left with the media and marketing person at Hollister, the oldest such business, went unanswered. Their mission statement includes the somewhat squirrelly "To make life more rewarding and dignified for those who use our products."

Defecating per abdomen is nothing new. As long as humans have been stabbing one another, intestines have been severed, and every so often the guts fuse with the skin and heal as an open portal. The phenomenon did not escape the notice of early medical practitioners, a few of whom took inspiration: "The lips of the intestine, so wounded, would sometimes quite unexpectedly adhere to the wound of the abdomen; and therefore it seemed no reason why we should not take hints from nature," wrote German surgeon Lorenz Heister in 1757.

Heister's colleagues dissuaded him, citing the cumbersome aftermath of the procedure. It wasn't until 1776[*] that a French surgeon, surname Pillore, undertook the inaugural cut, on a cancer patient with a hopelessly obstructed colon. "In this state," Pillore wrote, "that is to say, the patient having passed nothing from the bowel for more than a month, and his belly enlarging daily . . . I proposed to him that I should make for him an artificial anus." Pillore, too, consulted colleagues, "five or six"; all of them, like Heister's, deemed the aftereffects too inconvenient. Pillore ignored them, sharing Heister's view that it is surely better to part with one of the conveniences of life than to part with life itself. Sadly, Pillore's patient parted with both. He died a month later, though not of complications from his ostomy.

---

[*] Though the dates and location fit, there is no truth to the internet claim that Napoleon Bonaparte was an ostomate. Or, as some sites further claim, that he used a goatskin ostomy pouch. Or that the reason he appears with his hand inside his waistcoat in a famous Jacques-Louis David portrait is that he was hiding or adjusting his ostomy pouch. ("Hand-in-waistcoat" was a stock pose used by eighteenth-century portrait painters.) My source is Alessandro Lugli, lead author of "The Gastric Disease of Napoleon Bonaparte" and part of an international consortium of gastrointestinal pathologists who gathered on the occasion of the bicentenary of Napoleon's death to determine what had caused it. "I have read all the available autopsy reports," Lugli told me in an email, "but did not detect any reference that Napolean had a stoma." (The first ostomy receptacle on record was in fact a small leather pouch, dating from 1795, but it was worn by a farmer who, by a physician's report, "impaled himself on a wheat cart.")

By the 1800s, ostomies had become more common and more successful. The two earliest case reports on the medical journal database PubMed date from the 1820s and follow similar courses. We have Henry Baron, a forty-four-year-old bookkeeper from Manchester, England, and Mrs. White, a stout, ruddy widow from Bath, both seeking care for abdominal pain and severe constipation. Bowel obstructions are discovered and the standard indignities of the day unfold: "drastic purges" and turpentine enemas. Water propelled up the rectum by "powerful syringes." Repeated encounters with the rectal bougie, a sort of medical poker and stretcher. The list of laxatives prescribed by Mrs. White's doctor, Daniel Pring, runs from the predictable (castor oil) to the obscure (colocynth and elaterium), to a few so ridiculous-sounding that I pictured him turning to his children and asking them to make up a word: jalap, gamboge, scammony.

Eventually the day arrives when no further breach is possible. Mr. Baron, wrote his doctor, Richard Martland, "was now informed, the only chance for saving his life was by making an artificial anus . . . the inconveniences resulting from such operation, were candidly pointed out to him." Mrs. White, on the occasion of her twelfth day without a bowel movement, is presented with the same proposition. Both patients consented. Or rather, as Pring put it, "did not violently object." And so it went. "An opening was now made . . . and instantly a large quantity of liquid feces and wind escaped," Martland wrote. Both doctors were impressed by the force with which the matter was expelled, with Pring taking additional note of the "considerable distance" traveled.

The term "artificial anus" has writerly punch, but to me it's off-base. The hallmark, the bragging right, of a natural anus is its ability—thanks to the sphincter muscles that animate it—to hold things back. A surgical stoma affords no such control. As an anus, it's a flop.

It still is. The difference between then and now is the pouching system. The modern ostomate has more than a thousand discreet, waterproof, odor-proof pouches from which to choose. Henry Baron had a tin box. "This box," wrote Martland, "could be taken out and replaced at pleasure." Surely never before or since has the phrasing "at pleasure" been asked to bend so far. Mrs. White, for her part, was rigged like a Clydesdale, with leather straps and pads and a spring truss. She rarely ventured farther than the bedroom.

Ostomy surgeries today are often done so that the patient *can* leave the house. A significant percentage of patients have inflammatory bowel disease—mostly Crohn's or colitis. The intestinal tract, set upon by the immune system, becomes belligerent and erratic. It's not that these patients can't go; they're going too often and with too much urgency.

Mike Jones, a gastroenterologist I consulted for a previous book, described the typical patient. "They're on all these crazy immunosuppressive meds to try to get it under control, and it's not really under control. So they're still having diarrhea, and they're still soiling themselves, and they're spending a lot of money on medications that you probably don't want to take for the rest of your life anyway. At a certain point it's like, 'You know what? You just need to get your colon out.'"

Henry Baron lived another five years after his surgery. From time to time, he'd write to Martland. "He always," recalled the doctor in a sequel to the case report, "represented himself as far more comfortable than any one could conceive it possible to be in his situation."

Was Baron just being polite? Telling his surgeon what he thought he wanted to hear? Or was his situation not as onerous as Dr. Martland assumed? As we all assume. Of all the myths surrounding ostomy—you smell, you can't go swimming, you won't be able to bear children—is "hellish" the biggest myth of all?

The Arizona ostomy 5K is taking place in Scottsdale's Eldorado Park. I gave the Lyft driver the park address from the web page but did not scroll down to read "Parking Information," which has the precise location. It's a big park, and I now stand, confused, in front of the administration building. Two women sit at a table beside a sign offering free Bible study, but no one else is around. Presently a woman crosses the parking lot to join me. She pulls a small rolling suitcase with materials for the Colorectal Cancer Alliance table. We scan what we can see of the park. I fire off agitated texts to Ed Pfueller. The Bible ladies eye us hopefully. You would think that a Giant Inflatable Colon would be hard to miss, and I mention this.

"I'm the colon!" says my companion. "We own the colon."* Her colleague is bringing it in a separate car. She steps away to make a call, then turns back to me. "They're on the other side of the park." She offers a ride. My phone dings. It's Pfueller. "Ughhhhh," says the text. "Sorry."

A pale pink arch comes into view behind the mesquite trees a quarter-mile distant. It's the inflatable colon, our polestar. I head to the registration table. Representatives from ostomy supply companies have display tables set up alongside. The Hollister rep has pouches on a large silver ring, like upholstery samples. Choosing my empathy pouch is a crash course in ostomy life.

Very basically, there are one-piece pouches and two-piece "pouching systems." A one-piece is a pouch with a built-in adhesive ring. Stick it over the stoma and peel it off when it's

---

* You can own one too! The alliance sells inflatable colons in four sizes, equipped with blower, cord, storage bag, polyp, four Crohn's lesions, and diverticula. All but one of the colons take the form of a freestanding tunnel. The fourth, the Mini, is a flat cutaway piece of colon 4 feet by 6 feet. I considered buying one to use as an Aerobed, but they cost $4,000, and the polyps would be uncomfortable.

full. The Hollister rep compares it to a Post-it note. It's like a Post-it note the way an H-bomb is like a match head. The adhesive has medical-grade sticking power. A two-piece system has a pouch and a separate "wafer"—an adhesive-backed, ring-shaped patch that can be left on the skin around the stoma for a few days. In the systems I saw, the opening of the pouch snaps on and off the wafer, somewhat like a disposable vacuum cleaner bag.

Which to choose depends in part on where the stoma is. A one-piece is more likely to be used with a colostomy, as the stool is drier and more solid down toward the end of the tube (because it's been in there longer, so more water has been absorbed). Peristalsis pushes it out about as often as would happen with an intact colon and rectum, so the pouch needs changing just once a day or so. But if the stoma is up on the small intestine (for an ileostomy), liquidy waste seeps out more or less continuously, and the pouch would need changing a few times a day. Repeatedly pulling off a tenaciously adhered pouch can irritate the skin, so ileostomates usually opt for a two-piece system.

As with a natural rectum, the ability to release gas is critical. Just as a colon is in danger of a rupture should gas build up without an exit, a pouch can have what's colloquially known as a "blowout." To prevent this, pouches may have a built-in vent with an activated charcoal filter for absorbing odor. "Others prefer to burp it when it builds up," a sales rep says. "Like Tupperware." Burping can be done by popping on an Osto-EZ-Vent ("#1 preferred venting device") and pressing out the gas. *Fart* seems to me more appropriate than *burp*, for both ostomy pouches and Tupperware.*

Overwhelmed, I reach for one at random. "That's a large-bore, high-output bag," explains a sales rep. A woman standing

---

* The Tupperware people would prefer you use the verb *whisper*. Somewhere in the company's history, I learned on a visit to their headquarters years ago, it was decided that "burp" was not a "gracious" term.

beside me looks askance at my choice. "Very few people wear that." She holds it up to her shirt. "Look how long it hangs down. How unsexy is that?" The woman is Debra Adinolfi. She's with the Arizona UOAA affiliate. "I'm going to show you how little my bag is." She pulls up her shirt. Her abs are tan and toned. "See how little it is? I change it three or four times a day." At an ostomy gathering, there is no stigma. There's a *What are you wearing?* red-carpet fizz. It's excellent.

I go with an understated one-piece. Waiting for the race to start, I fall into conversation with a woman—let's call her Rebecca. She is young, perhaps early thirties, with lush long black hair, the kind that summons the word *tresses,* and equally lush false eyelashes. Rebecca had an ileostomy three years ago. Life is much improved since then. "Before, I couldn't do anything. Leaving the house made me anxious. Now I can travel, I can do what I want. Younger people are like, 'Oh, you have Crohn's? It's a tummyache, right?' I'm like, 'You have *no* idea.'"

A man standing nearby joins us. Rebecca asks how long it's been since his ostomy.

"Forty-nine years." Dave Rudzin developed ulcerative colitis as a toddler. At eighteen, his doctor proclaimed him out of other options. From 2010 to 2013, Rudzin was the president of UOAA.

I ask Dave and Rebecca if they're running. This seems like the wrong thing to say to a person with an ileostomy, but no one picks up on it. On National Ostomy Awareness Day, almost nothing is the wrong thing to say. Dave and Rebecca are both participating, but Dave, like me, is walking rather than running.

The race organizer blows a whistle and yells at the crowd to line up. Still yelling, she apologizes for yelling. "The sound system for the microphone isn't working!" It's that kind of event. I love it.

The organizer blows her whistle again, and Rebecca takes

off. Dave and I are quickly in last place, and that's just fine. The temperature is over 80, and it's barely 9:00 a.m.

I ask Dave about the procedure's persistent stigma. He says it's better today than back when he had his surgery, in part because of the candor of a younger generation and the reach of social media: TikTok ostomates, a lingerie model who has posed in her pouch, an uninhibited *Tosh.o* personality. Ignorance and prejudice remain the norm, however. People in Dave's subdivision have called him out for using the swimming pool. "I have a real issue with that," he says. His sunglasses are slipping, and he pushes them up the bridge of his nose. "I'll say, 'I showered before I got in. Can you say the same?'" There is an assumption that your hole is dirtier, your collection receptacle leakier. The gastroenterologist Mike Jones shot this one down. His girlfriend once worked at a country club where the average age of the members was around seventy. "She'd tell me, 'You wouldn't believe the number of discarded, shat-upon underwear in the men's locker room.'"

The double standard has been formalized at least once, in a 1987 fatwa. Mohamed Hanafy Ahmed, then the general manager of the Middle East branch of the ostomy supply company Convatec, had queried the Fatwa Commission of the Al-Azhar Complex of Islamic Research regarding ostomy pouches and ritual purification. "Is it possible," he asked, having explained the basics of ostomy, "for such a patient to pray while the pouch is carrying such excrements?" The commission chairperson replied that it would be acceptable provided the person first performs an absolution. They had to ask God's forgiveness. This seemed a little unfair. All of us, after all, routinely carry such excrements, in a pouch called the rectum.

What Dave said next was appalling. While he was at the helm of the UOAA, the Cincinnati Police Department announced plans for a public service campaign that would include an image of someone with an ostomy pouch—the aftereffect of a gunshot—to dispel the glamour of gangs and guns. "You're walking around with a colostomy bag and that's just not the way to get a girl's attention," remarked one officer, explaining the approach. The UOAA and the ostomy community petitioned for, and received, an apology.

Dave's most hardened scorn is for certain surgeons. While he was UOAA president, he would sometimes attend colorectal surgery conferences, where the organization had an exhibit table. "I would see surgeons walk by and do this." Dave turns to show me a look of derision and disgust. "I would get up and go after them. 'I saw what you did,' I'd say. 'And you're a colorectal surgeon? *Shame* on you.'"

A sprinkler head has left a shallow puddle on the path. I go around, but Dave walks straight through without adjusting his gait. It's so him: dogged and defiant, pushing forward, bending for nothing and no one.

Rebecca passes us, already on her second lap. She will go on to take first place in the Female Ostomate Division. She's barely sweating, her "skin bedewed with a gentle perspiration." (Old-timey medical loveliness courtesy of Dr. Richard Martland.) The eyelashes are holding strong. I think about how I once wore a pair, how strange they felt at first and how aware I was of their weight and presence—and how quickly those feelings faded. It occurs to me that I have not given a thought to the Hollister one-piece hanging off my abdomen.* Like wearing and changing Depends or a maxi pad, it is, I suppose, as Dave said to me at some point, "just something to get used to."

Part of the aversion, Dave believes, is to surgical disfigurement: the hole where no hole is meant to be. Whether it's that,

---

* Writing this now, months later, I see that I overlooked the bit in Ed Pfueller's original email to me, which reads, "filled with the fluid of your choice." My empathy bag was empty! I've since co-opted the phrase for use in dealing with emotionally clueless people. *Yeah, his empathy bag is pretty much empty.*

or the bag or its contents, or more likely all of it, life with an ostomy can only be judged side by side with the patient's life without one. The Fecal Incontinence Quality of Life Scale* has a few ways to describe that. *I try to stay near a restroom as much as possible. I am afraid to go out. The possibility of bowel accidents is always on my mind.*

Back at the park ramada, hot dogs and hamburgers are cooking. A cover band plays Tom Petty. People hang out for another hour, enjoying each other's company, until the Colorectal Cancer Alliance starts deflating their colon and it's time to go home.

If you, like me, have ever had cause to search the PubMed database for research on artificial anal sphincters, then you, like me, might have noticed one university standing tall. The sphincteric wizards of Shanghai Jiao Tong University have published close to a dozen papers, laying claim to an impressive array of backdoor gadgetry: A wireless hydraulic-electric anal sphincter with communication subsystem. A puborectalis-like artificial anal sphincter that replicates rectal perception. An intelligent, remote-controlled artificial anal sphincter with external telemetry unit and transcutaneous energy transfer system. Shanghai engineers have developed sphincters with anal pressure sensors, medical micropumps, real-time feces monitoring.

---

* Not to be confused with the St. Mark's Incontinence Score. Intrigued by the possibility of a patron saint of fecal incontinence, I turned to the internet. There I learned that the title of the scale comes not from the saint but from the university where it was created. Mark is the patron saint of a confusing assortment of individuals, including pharmacists, lions, secretaries, and "people dealing with insect bites," but not ostomates. The closest we have to a patron saint of fecal incontinence is Saint Polycarp, patron saint of diarrhea and dysentery. Boosters of St. Bonaventure tried to capture the market by noting that their man was in fact patron saint of all bowel disorders, but with patron saints, I feel, specificity rules. My vote goes to St. Polycarp, with St. Elmo, patron saint of abdominal pain (and sailors), as runner-up.

There are biocompatible sphincters and sphincters crafted from shape-memory alloy.

Alas, none appears to have made it beyond the lab. I could find no published clinical trial results. One device was tested on a pig intestine, but it wasn't clear to me that a pig was attached. I began to suspect the influence of a single, determined professor, hopeful and incontinent, who every year tasks his students with the design of an artificial anal sphincter. The suspicion has been neither confirmed nor countered, as inquiries to the researchers and the engineering school administration went unanswered.

Here in the United States, at least one artificial anal sphincter has been trialed and approved by the FDA. The manually operated Acticon Neosphincter is a surgically implanted cuff which, when fluid is pumped into it from a receptacle in front of the bladder, bolsters the natural stranglehold of the internal anal sphincter. When stretch receptors in the rectum signal that it's full and the person wishes to empty it, they squeeze a bulb implanted in the labia or, lacking those, the scrotum. This drains the fluid from the cuff back into the receptacle and relaxes the hold.

In a five-year follow-up study of twenty-eight Neosphincter users, just fourteen still had the device in place, and only three reported good results. Eight no longer activated the pump, citing obstructed defecation. Mike Jones summed up the basic problem: Artificial anal sphincters either don't clench tightly enough or they don't "relax" sufficiently. "So you walk this fine line," he said, "between providing enough resting pressure to keep stuff from falling out, but not so much that it's an insurmountable resistance, so you can't push and defecate." So that you don't swap incontinence for constipation.

Jones had heard that better results were possible using magnets to hold the sphincter shut. I wrote to the lead author of a paper on one such a device. "You recently contacted us about the Fenix system," said the public affairs person who intercepted my email. "Please know that it has been discontinued and is no longer on the market."

"I'll tell you, none of them are used a lot," said gastroenterologist Peter Cataldo when we later spoke by phone. (Also probably not used a lot: Cataldo's academic title. Deep breath! The Samuel B. and Michelle D. Labow Green & Gold Professor of Colon & Rectal Surgery at the Larner College of Medicine at the University of Vermont.) Cataldo says it's very rare that someone gets an implantable sphincter. "It's foreign material, and it's very near your anal mucosa. So if there's an inflammatory reaction to the material, and the device erodes into the mucosa, now you have a piece of foreign material that's exposed to bacteria from the colon. And that's a big fat disaster." A quarter of the subjects followed in the aforementioned Neosphincter paper had infections serious enough to require an added round of surgery.

So why do these things get approved? For one thing, the safety and efficacy trials that the FDA requires for devices can be smaller—meaning fewer subjects—and shorter than they are for trials of new drugs. With a smaller number of subjects and a brief time span, complications may be underestimated or missed altogether. And once one device is approved, an inventor of a newer device in the same category need not conduct trials at all, provided they can show that theirs is "substantially equivalent," whatever that means.

Rather than construct a sphincter from artificial components, might it be possible to grow one? Aren't we growing and 3D printing everything in labs these days? The Wikipedia entry for "internal anal sphincter" states that in 2011, the Wake Forest University School of Medicine announced that "the first bioengineered, functional anal sphincters had been constructed in a laboratory made from muscle and nerve cells." If true, it is an astonishing achievement, as is that lab built out of muscle and nerve cells.

I found a couple of the papers. Indeed, researchers at the

Wake Forest Institute for Regenerative Medicine (in 2017, not 2011) wrote that they'd grown rabbit "biosphincters" from cultured muscle and nervous system stem cells taken from the animals and seeded onto a sphincter-shaped mold. The resulting structures, which resembled the rolled ends of condoms, were implanted in ten rabbits that had been surgically rendered incontinent. Compared with controls, the animals with the lab-grown equipment had better pellet retention and kept their tushes spotless and white. Whereas the untreated rabbits left messy droppings all over the cage and showed "fecal staining" of their fur. Close-ups of the anuses and the cage floors were included, under the heading "Rabbit Hygiene."

Encouraged, the researchers moved on to nonhuman primates. Their 2019 paper brought astounding news: the lab-grown sphincters "resolved fecal soiling" and showed healthy resting muscle pressure. Incredibly, the monkeys were reported to even have a working recto-anal inhibitory reflex, or RAIR. The RAIR is what opens the inner anal sphincter when the rectum is full and ushers the material into the anal canal, ready for offloading. Most impressively, it tells you what that material is: solid, liquid, or gas. The RAIR, according to its very own Wikipedia entry, "allows for voluntary flatulation* to occur without also eliminating solid waste." That is to say, farting rather than sharting.

It's been more than five years now. Are there people at this moment walking around with functional lab-grown anal sphincters? I emailed a Wake Forest public affairs person. "The researcher has retired," she tried. And no one else

---

* Okay, *flatulation* is not a word. Not according to the *American Heritage Dictionary*, not according to *Merriam-Webster*. Does the textbook *Gastrointestinal Physiology*, 2nd edition, actually, as the Wikipedia source list suggests, employ the word *flatulation*? It does not. Wikipedia, unlike your rectum, is unable to tell when it's full of shit.

carried on the work? Fecal incontinence affects one in twelve American adults!

"It never got to human clinical trials," the public affairs woman said, "and I can't speak to why."

Cataldo spoke to it. "That stuff is not ready for prime time. It may happen one day, but it's a long way away."

At the moment (it's mid-2024 as I write this), the only human cell–based body part to make it to a human (FDA Phase 1) clinical trial is a 3D bioprinted ear. Actually, it's just the outer part, the pinna. Bioprinting guru Adam Feinberg, who leads the Regenerative Biomaterials and Therapeutics Group at Carnegie Mellon University, respectfully described it as low-hanging fruit—in that the outside of an ear is minimally functional and doesn't need to bear weight. The team in his lab is reaching for fruit higher up the tree. In the course of a half-hour phone call, Feinberg mentioned projects devoted to the eventual printing of hearts, livers, pancreatic islets, leg muscle, and tendons. It seemed the lab was gearing up to print, bit by bit, an entire person. (And something for it to eat; they're collaborating with a colleague on printable Wagyu beef.) I have a hard time getting my Epson ST-M1000 to spit out a twenty-page chapter, in simple black-and-white, without a paper jam. How do you print a whole organ? How close are we to doing it? Who do you call when there's a liver jam?

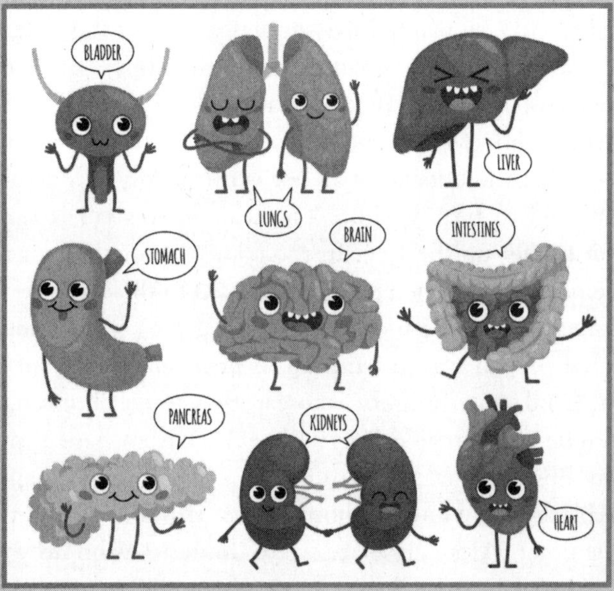
BLADDER
LUNGS
LIVER
STOMACH
BRAIN
INTESTINES
PANCREAS
KIDNEYS
HEART

13

Out of Ink

How to Print a Human

The first "ghost heart" showed up online around 2005. In photographs, it has every appearance of being a normal heart, except that it's white. This is because the cells have been washed away, leaving only collagen and other pallid human support matter—the stuff known collectively as the extracellular matrix. *Wash* is a technically apt verb, as the deed is done with detergent. Detergent, be it Dawn or Triton X-100, breaks down lipids and makes them easy to rinse away. In addition to supplying greasy residue on plates and pans, lipids make up the outer membranes of our cells. So if you infuse an organ with detergent—that is, you pump it in via the vascular system, much as a mortician distributes embalming fluid—it will handily dissolve these cells. In theory, you could decell an entire body, leaving behind a pale husk, a sort of humanoid sheath with a skein of empty blood vessels. It's a mystery to me that no horror film has yet featured a scientist turned serial killer who decellularizes his victims.

Decellularization is the "decell" portion of a bioengineering process known as decell/recell. Collagen has an appeal for would-be organ builders. It has no live cells, and thus none of the cell surface proteins that prompt the immune system to go on the attack. The body will accept foreign collagen—from cows, pigs, most any mammal—without much fuss. Once you've decelled, say, a pig organ, you would use the same network of capillaries to recell the scaffold that remains—to deliver millions of the patient's own lab-cultured cells. In this way, the thinking went, it might be possible to construct whole replacement organs, ones that require no immunosuppression.

Decell/recell inspired the career of Adam Feinberg. "It was such an intriguing idea," he said when I visited his lab at Carnegie Mellon one spring morning in 2024. "This notion that you could take a pig heart, remove the cells and keep the collagen, and then replace the pig cells with human cells." He cracks a knuckle. "Turns out that's very difficult."

There's a problem. When detergent dissolves cells, the rubble takes the form of molecules, which, because they're extremely tiny, are easy to flush away. Whole cells are ten thousand times larger than molecules. Feinberg describes it as the difference between running a 5K and running around the earth. So when you try to reseed a scaffold by pumping in the patient's intact cells, they're too big to pass through the capillary walls. "It's almost impossible to get the cells back." Feinberg has a vibe of fitness and calm confidence, partly from his posture and build, but also from his voice, a splendid broad-shouldered baritone that carries without effort. A voice for the stage. It's like having Macbeth explain regenerative medicine for you.

"And even if you could flow them in, how do they get to the right spots?" The liver, for instance, is built of five major types of cells. How would they know where to hop off the ferry? The answer, for now at least, lies in 3D bioprinting: building the cellular tissues and the collagen scaffold together, as you go, in thin cross sections. Imagine printing a hardboiled egg layer by layer, the printer switching materials as it moves along: shell, white, white, white, white, yolk, yolk, white, white, white, shell. For human bioprinting, one ink would contain the extracellular matrix, mainly collagen, and others would contain live cells. Using 3D data from a patient's MRIs, the printer would lay down the various inks, switching among two or more extruder heads as it travels the plane it's printing. (Other 3D bioprinting processes exist, but for the sake of simplicity, we'll stick to extrusion.)

A problem here, too: Organs are malleable—flimsy, even. An extrusion 3D printer builds what it's building from the bottom up, the first layers serving as the base for the layers to come. To facilitate this, plastic inks are made to solidify within seconds. I own a 3D printed facsimile of a human rectum, made of hard plastic. You could not print an actual rectum this way, any more than you could build a cathedral out of tofu. The farther up you build, the floppier the structure becomes.

Feinberg invented a process called FRESH,* whereby the printer extrudes the construct—that is, the thing it's printing—inside a gel "support bath." He has enlisted a grad student, Caner Dikyol, to demonstrate. Caner (pronounced *cha-NAIR*) is gangly and, despite his obvious commitment to the work, endearingly goofy. He selects a needle-thin nozzle and attaches it to the printer's extruder. Underneath he slides a Petri dish the size of an antique pocket watch, which is filled with a pale amber gel: the support bath. The main ingredient, Feinberg chimes in, is basically the main ingredient in hair gel.

I glance at Feinberg's hair, which is cut close to his head but has the sheen of product. He intercepts the glance. "No, I have never tried it."

We chat for a while as the printer does its thing. Caner asks why I own a 3D printed rectum, a not unreasonable question. I explain that it was a gift from a radiologist who had interviewed me onstage about a book that includes a rectum chapter. "He printed it with a base, so I can stand it on a shelf. It's bright red and—"

The giddy exuberance that is Caner Dikyol dims abruptly.

---

* Freeform Reversible Embedding of Suspended Hydrogels. Since you asked, or didn't.

"We were thinking to give you an artery." He indicates the Petri dish, the little worm materializing in the costly hair gel.

"But this is so much better," I lie. "It's *collagen!*"

Half an hour later, Caner slides the Petri dish out from under the printer head and holds it up for me to see—the artery suspended like fruit in a serving of Jell-O. He walks it over to an incubator for the final step. The support bath is temperature-sensitive; heat causes the gel to melt away.

While my souvenir warms, Caner and Feinberg and I wander over to visit a postdoctoral student, Ali Asghari Adib, who is printing liver cells. (Actually, liver tumor cells, a commercially available cell line[*] used in liver-related research.) This printer is larger and fancier. It can switch among up to four extruders, and does so using the same no-friction electromagnetic technology that makes possible the fast, smooth ride of a bullet train. Two extruders are going, one for the collagen ink and one for the ink with the live cells. Feinberg likens it to printing a brochure with different color inks.

The nozzle that extrudes the bio-ink is about the diameter of a human hair. You can push a liquid through an opening that small, but neither end product here, not the collagen or the live cell ink, is liquid. Here's what Feinberg came up with. The collagen ink is acidified, which keeps it in liquid form while it's being extruded. When the ink hits the support bath, the pH changes to neutral. For reasons we need not dive into here, this prompts the liquid collagen to assemble itself back into fibers.

This ingenious bit of laboratory alchemy won't work with a bio-ink of live cells, however. The acidity would kill them.

---

[*] Specifically, HepG2—an "immortal" cell line that has been contributing to liver research since 1975, when it was isolated from the tumor of a fifteen-year-old Argentine boy who has not, unlike Henrietta Lacks and her tumor, been the focus of a popular science book that has sold more copies than there are cells in a $500 cryovial of HepG2.

In this case, Feinberg took inspiration from the body's clotting mechanism. The live cell bio-ink contains fibrinogen, which turns to fibrin, the stuff of clots, as we learned and forgot in chapter 4, when it's exposed to the enzyme thrombin. So Feinberg's support bath contains thrombin, which causes the fibrinogen-spiked ink to solidify on contact—to clot, basically—once it's printed.

Here's another difference between printing a brochure and printing living tissue: brochures don't need to be fed. Human cells get what they need to survive—nutrients, oxygen, waste disposal services—via capillaries. You can't go more than about the diameter of a hair without encountering a capillary. They're everywhere. You can't print all those little tubes. The printer would be changing heads every other second. Fortunately, the human body excels at building capillaries on its own. "That's what happens when we get fatter or build muscle," Feinberg says. So if you're printing tissue, you cross your fingers that it will sprout its own vasculature. (Feinberg encourages this by adding growth factors to his inks.) The body is not good at—or even willing to try—building large vessels. Consider the aorta, a multilayered blood vessel so large it has interior capillaries. It's vasculature that has its own vasculature.

"We're in this period," Feinberg says, "of trying to understand: How much do we have to build versus how much can we get the body to do some of the job for us?"

The cells Ali has just printed, I'm told, will get to work in a few days, dutifully secreting albumin for no one. I find astounding the extent to which a cell, independent of an organ or even a body, will proceed to do the job it's born to do. Liver and pancreas cells make and secrete hormones. Heart muscle cells beat. Earlier, I was given a microscope view of thousands of heart cells that had, with no prompting from their human caretakers, begun beating in unison. Cardiomyocytes have an intrinsic behavior whereby when two of them touch, Feinberg explained, they synchronize beats by opening a tiny passage between themselves. Soon a third connects to the pair and takes up the beat, and so on, until the herd is drumming so vigorously it occasionally catches air, losing contact with the base of its dish.

The liver bit is being printed as a mixture of extracellular matrix, cells, and proteins that support growth. Ali will watch to see what, if anything, self-assembles. Some of the constructs have grown capillaries, but it hasn't happened consistently. "So now we are troubleshooting," he says.

Why not start with something easier, I ask them. Maybe cartilage. It's basically one type of cell, with no nerves and no vasculature. On an earlier reporting trip, I encountered a transplant surgeon who had enthused about the future of the 3D bioprinted meniscus, a fibrous pad between the bones of the knee. "Pop it in, and you're good to go," he'd said.

But not very far, according to Feinberg. "No one's figured out how to make it as strong and tough as normal cartilage," he says. In studies, 3D bioprinted replacement cartilage has been shown to be helpful, but no more so than current treatments on offer. "So, yeah, you can make a thing that looks like a meniscus, with the right cells, but you can't make it with the right properties, the same durability," Feinberg says, quickly adding, "yet."

Everything is more complicated than you think it's going to be. It took Feinberg's team two full years to figure out how to print collagen that can be sutured in place without the stitches ripping through. It wasn't just the mechanical properties of the collagen filaments that mattered, but how they were aligned in the printing—their direction relative to the pull of the sutures.

In 2006, Wake Forest sent out a splashy news release: "Wake Forest Physician Reports First Human Recipients of Laboratory-Grown Organs." Cells were cultured from patients' biopsies and pipetted onto collagen scaffolds and allowed to grow for seven weeks. The resulting structures were stitched into the malfunctioning bladders of children with inherited anomalies. I had heard about this, in general terms, from an acquaintance who runs a stem cell lab at Temple University. One of the kids, she said when we spoke, had just graduated college with his Wake Forest bladder still in place. ("It survived all the beer!")

I mention this to Caner, how these entities supposedly functioned like natural bladders.

"Supposedly." Caner isn't convinced. A working bladder relies on communication between muscle and nerves in order to sense fullness and empty itself. "There was not this muscular function," Caner says.

Functional muscle is an extremely challenging tissue to print. Within the musculoskeletal system, for example, muscles need nerves to carry out their work. A heart can do without, as hormones help control its function, but a limb takes orders from the brain. Without nerves, leg muscle is just meat. "Nerves are a separate issue that we're not yet tackling," Feinberg allows. No one in the field is printing working human nerves. They're challenging, in that they have a limited capacity for regrowth.

To function, any variety of muscle cells needs to be printed in a precise manner that serves its calling. That is, the cells have to be appropriately aligned. In the deltoid, for instance, cells are arranged in a fan shape; this is partly what gives the shoulder its impressively broad range of motion. The cells of the hamstring, a muscle at the back of the thigh, are laid out in parallel, enabling the quick contractions of locomotion. Cells that make up sphincters are aligned in rings, with the rings themselves laid out in a loop. When the rings contract in unison, the loop shrinks, applying the needed chokehold.

Caner brings me to the workstation of Maria Stang, a postdoctoral researcher whose days currently revolve around printing the various architectures of human muscle. Stang's own architecture is exquisitely aligned. Everyone I've thus far met here is nice-looking, but Maria Stang is ridiculous. If I'm a 5K, she's a run around the Earth.

Stang has been working on printing the trickiest architecture of all: heart muscle. Cardiomyocytes are arranged in a helix shape around the heart's chambers. So as the organ beats, it not only squeezes but also twists slightly. Feinberg compares it to wringing a wet towel. This serves to maximize the volume of blood that's pumped with each beat.

Alignment is especially important with heart muscle, because without it, the cells' electrical impulses fire arrhythmically, and nothing, pumping-wise, is achieved. Stang has succeeded in creating heart constructs that beat in a coordinated manner. That's huge. Unlike the beating heart cells I saw earlier under the microscope—a single layer of a single type of cell—Stang's constructs are one millimeter thick. (There are no capillaries; for now the cells are printed with a low density that allows nutrients and oxygen to diffuse through them.) Like actual heart tissue, they're three-dimensional structures that incorporate struts of collagen.

"I'm going to show you the tissue which is contracting the best." With Feinberg tagging along, Stang leads me to an adjoining room. She readies a slide and slides it onto a microscope platform, then steps away to let me look. There is an awkward quiet. I can't see any movement. "It's beating very slowly," Stang says. "Maybe thirty beats per minute."

"Hm."

Feinberg leans in. "Is it in focus for you?" It is.

"Watch the sides," Stang directs. "See how they squeeze in?"

"Oh yes." It's subtle. "Yes, I think so." My husband has a friend, Dale, whom we once took to watch a meteor shower, because he'd never seen one. The word *shower* had led Dale to expect legions of stars streaming across the sky at once. So of course he was underwhelmed. Right now I am Maria Stang's Dale. I've read so many overblown press releases and news bits about 3D bioprinting that the reality, this amazing, near-godly little chunk of working human heart, has failed to spark the wonderment it unquestionably deserves.

Stang takes it in stride. "Obviously there's more work to be done."

I ask Feinberg when he thinks medical science will arrive at the point of implanting entire functional bioprinted organs in patients. If we use the analogy of airplane flight, he puts things somewhere around the Wright brothers stage. "Of course, we don't want a plane that goes thirty feet down the field. We want a plane that can fly around all day."

And how far off is that? A decade plus, Feinberg says.

For medical science, that's actually a brisk turnaround. (In an earlier phone conversation, Feinberg equated "a decade or two" with "pretty quickly.") He adds that he thought it could easily happen far sooner. "Because we keep coming up with new things." Just twenty years ago, he points out, there was no gene editing, no CRISPR. "Plus AI is going to accelerate, and that's going to change what's possible."

I pose the same question now to Jaci Bliley, a senior postdoctoral fellow with wispy waist-length blonde hair and what corny Irish lyrics call smiling eyes. Bliley has just joined us in the microscope room. Two to three decades is her estimate. Like Feinberg, she says she's surprised at how fast things are moving. She offers the example of some stand-alone beating heart ventricles, little tubular constructs that she printed as part of her PhD research. "That was 2019," she says. "Now we're putting them into mice and they're surviving. After six months they're still alive and beating."

They're keeping mice alive with 3D bioprinted heart chambers?!

Not exactly. The mice kept their natural hearts. The printed ventricles lack valves, which are needed to keep blood from backflowing, so it moves in the direction you need it to, regardless of the pull of gravity. When Bliley's mouse ventricles pump, the blood shoots out unhelpfully in either direction. Collagen valves will be incorporated soon. The lab has already printed them, and they function perfectly.

I guess because I've had a glimpse now of some of the hurdles to be cleared, it's hard for me to imagine the day when organs are being routinely printed, like car parts, and installed. I ask Bliley if she ever feels daunted by the amount of work that remains or discouraged when experiments fail. She shakes her head. Her hoop earrings sway. She wears a look that suggests she's just a little bit disappointed in me. "It's never a failure. You've learned something. It's all progress. It's *exciting*." From Bliley's expression it's clear she isn't upselling the positive just because Feinberg is standing here or the university public affairs person would like her to. "For the rest of my life," she'd said earlier, while we were walking across campus, "I can't imagine doing anything else."

As the beneficiaries of this kind of passion and dedication, we owe our scientists a lot. We owe them gratitude, awe, respect. Mostly what they want from us, of course, is a little more funding.

Money is an ongoing hurdle. Feinberg estimates it would cost around $100,000 to print one heart—approximately the lab's budget for a semester.* Heart cells are unique in that they don't replicate. You can't buy a cryovial and expand them out. You need to create them from commercially available induced

---

* Like most, Feinberg's lab relies on grants. Along with collaborators at the Mayo Clinic, they recently applied to a new federal research funding agency, Advanced Research Projects Agency for Health, or ARPA-H, which has earmarked one to two hundred million dollars for organ bioprinting. Why ARPA-H and not HARPA, to match our other catchily acronymed advanced research agencies—DARPA (defense), BARDA (biotech), and IARPA (intelligence)? I'm guessing the shift toward blah began with ARPA-E, which, had it been EARPA, would have suggested an advanced research projects agency devoted not to energy but to ears.

pluripotent stem cells (iPS cells). This entails a time-consuming process of differentiating them and feeding and caring for them. "They're like infants," Bliley says, more with fondness than as a complaint. "On weekends we're here. On Christmas we're here. I'm a cell mom."

The cells that would one day be used to print actual hearts for transplant may be yet more expensive. Until such time as hypoimmune, or "stealth," iPS cells are available, one would need to work with the patient's own cells in order to prevent rejection. As we'll see, this is a costly and time-consuming prospect. It's done by regressing the patient's cells—typically something easily accessed, like blood cells—into an earlier, undifferentiated (pluripotent) version of themselves. The regressed iPS cells are then coaxed into becoming cells of whatever variety is called for to build the replacement tissue.

As with much of regenerative medicine, smaller is easier. A ventricle is easier than a heart; a mouse ventricle, easier still. One of the companies furthest along has set its sights on the smallest organ of all: the human hair follicle. Long before regenerative medicine grows a heart or reverses liver failure, it may well conquer baldness.

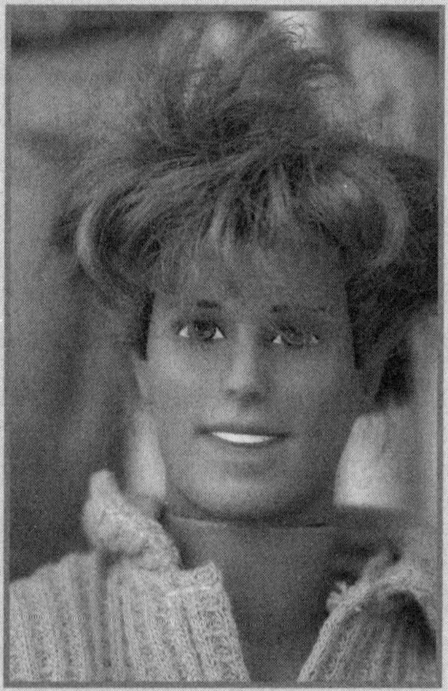

14
Shaft

Hair Transplants Through the Ages

In the late 1950s, exact date unknown, a dermatologist named Norman Orentreich picked up a circular punch and cored out four flesh pellets from the back of a man's head. The punch, a sort of tubular scalpel, is mainly used for taking biopsies, but this man had no suspicious growths. He was merely losing his hair. Orentreich then made four short cuts in the skin at the top of the man's head, took up the pellets with tweezers, and planted them. As one might a box of tulip bulbs. A few months later, the transplants sprouted hairs. As did transplants on fifty-one more balding subjects. Orentreich watched the hairs grow for the next two years, and then, with what must have been considerable excitement, he presented his findings at a professional conference.

The audience booed. Everyone knew, or thought they knew: You can't restore hair where a scalp is genetically inclined to lose it. If the hairs haven't fallen out yet, colleagues scoffed, they soon would. And yet they didn't. Orentreich had demonstrated a phenomenon he named "donor dominance." That is, when you transplant follicles, the hairs retain the characteristics of their homeland. Hair follicles from the sides and back of the scalp lie outside the pattern of male-pattern baldness. They don't go dormant, and they retain that golden attribute when they're transplanted up top.

In a 2009 letter to the editor in the *Journal of Dermatology*, a pair of dermatology professors at the medical school of Osaka University noted that although Norman Orentreich is "recognized as the 'father' of hair restoration surgery," he was not the first to transplant follicles. Twenty years earlier, Japanese physician Shoji Okuda documented in his journals the successful transplantation of hundreds of "hairy columns." (Elsewhere translated as "plugs." Even the vernacular is Okuda's to claim.) Out of respect for Orentreich, the Osakans simply asked readers to consider Okuda "'another father' of hair restoration surgery."

Because cosmetic surgery carried a stigma at the time, Okuda focused on transplants for sufferers of some of the rarer forms of baldness—pubic alopecia* and alopecia areata, wherein the immune system attacks a person's follicles—and for scars on the scalp and eyebrows. Unlike Orentreich, Okuda considered any hairy plot of skin to be a potential donor site. He experimented exuberantly. Armpit hairs were transplanted to heads, head hairs to pubic areas, pubic hairs to eyebrows. "Using this method," he wrote in his journal, "very easily I could transplant living hair any place."

---

* A word about pubic hair wigs. Merkins were supposedly used by sex workers in bygone centuries to conceal that they'd shaved their pubic hair—a sign of lice. But a false thatch held in place by spirit gum or suspended on a string seems unlikely to fool anyone in an intimate encounter. I asked wigmaker and amateur adornment historian Loryn Pretorius whether merkins were really a thing. Neither she nor I could find one in the holdings of relevant archives and museums. And there's no information, she said, in any of her wigmaking books from the eighteenth and nineteenth centuries. The evidence put forth for their existence takes the form of a handful of literary references in the form of jokes or insults. (The oft-used online photo purporting to show an 1820s merkin vendor—J. G. Paine & Sons, "serving the discerning pudendum"—was staged.) However, contemporary merkins verifiably exist. A colleague of Pretorius's has made them for actors seeking whatever slim modesty can be supplied by wearing a sort of hair-fronted thong during full-frontal nude scenes. Pretorius herself has not, though she can lay claim to a pair of sideburn wigs for Dustin Hoffman.

Without intending to, Okuda had supplied a solution to the lasting challenge of hair transplantation for male-pattern baldness: supplies are limited. As male-pattern baldness progresses, additional transplants are needed, lest a swath of bare scalp develop behind the original transplants. Okuda's multisourcing legacy lives on. Surgeons at, by way of random example, the Manhattan hair restoration clinic City Facial Plastics, will transplant hair to the head from the chest, armpit, beard, pubic area, abdomen, back, arms, or legs. (The reverse is also on offer: "scalp to body transplants," for patients who want hairier chests or fuller pubic hair.)

Though Okuda undertook no transplants for male-pattern baldness, his journal mentions a single purely cosmetic transplant: follicles taken from a man's scalp to fill in his "sparse mustache." In a photo that accompanies a paper by Kenichiro Imagawa and Shigeki Inui in *Hair Transplant Forum International,* Okuda sports a remarkably lush "toothbrush"* mustache. I wondered if the beneficiary of the mysterious solitary mustache transplant had been Okuda himself. "Absolutely impossible to do on himself," came Imagawa's emailed reply. I apologized for my impertinence, and I apologize to you, reader, for this diversion, this unruly strand in my tidy narrative hairline.

In 2009, Imagawa and a colleague made a pilgrimage to Okuda's clinic and warehouses, which still stand, their contents preserved "just like in the tomb of Tutankhamen." They chatted with Okuda's grandsons, who recalled the "many before/after pictures of hair surgery in the hall of the clinic that made

---

* Better known today as a Hitler mustache. Though Hitler didn't start the fashion. He ended it. The toothbrush mustache first appeared in the U.S. as early as the late 1800s and remained popular worldwide until World War II—when the association with Hitler abruptly tanked it. Hitler wasn't the only prominent Nazi to adopt the style. Heinrich Himmler and Ernst Röhm both sported toothbrush mustaches, as did, appropriately, Hitler's dentist, Hugo Blaschke.

them feel sick in their childhood." (The grandsons later found Okuda's original punches and gifted them to the two visitors.) Before leaving to drive home, the men paid their respects at Okuda's grave. "With bowed heads," Imagawa wrote, "[we] reported to him the story of modern hair transplantation." Had you been visiting the cemetery that day, you might well have been witness to a memorable scene: two men speaking to a grave marker about scalp extenders and strip harvesting and the Juri rotational flap.

In the fifteen years since, the story has undergone a significant plot twist. If things go where they appear to be heading, people won't be limited to transplanting the follicles they were born with. They'll be able to grow—off-head—crops of new follicles derived from their own cells. This is the dawn of unlimited hair. And by extension, the dawn of unlimited cell regeneration. If we can grow replacement follicles, it stands to reason one day we'll be growing replacement everything. Hair seemed simplest, most likely to be something I could, with my nonexistent regenerative medicine background, understand. And so I'm paying a visit to Stemson Therapeutics, a San Diego company that is furthest along in the game.

Stemson's venture began with basic follicle research. They needed to observe in detail the journey taken by the cellular building blocks of new follicles—dermal papilla cells and keratinocytes. Without documenting the coming together of the two types and the hair's travels up and out of the skin, they could not be sure their lab-generated follicles were hitting the same landmarks as follicles growing au naturel.

That is where plastic surgeon Richard Chaffoo comes in. Chaffoo is a veteran hair transplanter whose office is two miles down the road from Stemson. When patients arrive for a transplant, Chaffoo asks them whether they'd be willing to donate some follicles to science. If they say yes, a dozen or so are placed on ice in a cooler and driven to Stemson's lab. Today the follicles will come from my head.

A hair transplant is an all-day job. Over the course of a morning, follicles—or more accurately, follicular units, which typically hold two or three follicles—are harvested. One to two thousand of them are removed one by one from the back or the sides of the scalp. After a break for lunch, they're implanted. Again, one by one. You can transplant a kidney in half the time it takes to finish a hair transplant.

By my reckoning, Richard Chaffoo has transplanted upward of fifty million hairs in his thirty-five years bent over strangers' heads. He seems, understandably, a little bit done with it. He's been dabbling in other pursuits. He's working on a book. When a Stemson founder approached him about helping out, he happily signed on.

I arrive at eight on a warm November (it's San Diego) morning. Chaffoo is at his desk in a sunny office down the hall from the operating area. Are you old enough to picture Peter O'Toole? That's the general look of Richard Chaffoo: long face, blue eyes, fair hair. Outstanding cheekbones. The office is nicely appointed in the style of some recently past decade: rugs then known as Oriental, ball-and-claw foot furniture legs. Chaffoo's computer monitor rests on two volumes of *Pediatric Dermatology*, and dear God there shouldn't be enough for two volumes.

What draws a surgeon to hair? For Chaffoo, it was the artistry of it. You are, he says, designing a hairline. *Line* is a misnomer. "It's a zone of transition. It can't be too straight or too abrupt. You want some irregularity." This is achieved by implanting follicles in a slightly meandering manner, a technique called "snail-trailing." "And you don't want to give a fifty-year-old the hairline of a twenty-year-old," Chaffoo says. The name

Elon Musk comes up. Chaffoo lowers his voice. He leans forward in his chair. The armrest bangs the desktop. "No one's hair line looks like that." It's too straight, too thick. Chaffoo guesses Musk has had multiple procedures. "He's going to run out of hairs."

As per Okuda's work, Musk could, in a pinch, fill in with body hair. He doesn't look like a hairy individual, but perhaps a little something from the armpit or pubes. He'd need to stick to a short hairstyle, as body hair doesn't remain in the growing (anagen) phase much longer than a month. (The reverse has been an issue for women who get transplants of scalp hair to address pubic thinning. "The patient herself may have to cut her hair once every 1 or 2 months," advise the authors of a *Dermatological Surgery* paper on pubic hair restoration.)*

Although Chaffoo once treated a woman with pubic alopecia, he has not undertaken the reverse variety of transplant. "It'd be kinky and coarse," he says with some distaste. And then, in case I'm not getting it: "It's pubic hair on your head."

"Donor dominance!"

"Donor dominance, yes."

Some men are fine with it. In 2016, Redondo Beach dermatologist Sanusi Umar carried out a study on body hair transplants and patient satisfaction. He transplanted 823 pubic hair follicular units (about 1,400 pubic hairs) to patients' heads, with a mean overall satisfaction rate of 8 on a scale of 1 to 10. This despite

---

* Surely a smaller market in the current era of pubic denuding, a practice now commonplace among men as well. Did men ever seek out transplants for scrotal alopecia? "The level of concern by men regarding the development of hair loss from their scrotum and the demand for treatment . . . remains to be determined," writes the author of the journal paper "Scrotal Rejuvenation." For those who question the existence of scrotal alopecia, the author includes a photograph (Figure 1: Scrotal Alopecia and Scrotal Laxity in a 93-Year-Old Man) depicting "nearly complete absence of hair" and, not that you asked, "low-hanging scrotum that extend[s] to the mid-thigh."

unique grooming challenges: to wit, the "relatively wirier" hair being "difficult to style."

I tell Chaffoo I'd like to experience donor dominance, to marvel at the strangeness of a few strands of long, flowing head hair growing on, say, my leg. He's not sure it'll work. There's less blood flow to the calf than to the head, and the skin is thinner. True, I say, but I've looked into it and the odds of success are decent. According to a 2002 *Dermatological Surgery* paper I've brought along, the survival rate of scalp hairs transplanted to the lower leg is 60.2 percent. (I have no clue why anyone needed to know this.) Chaffoo agrees to do it, if only because it's easier to move a follicle than to argue with a nitwit.

He brings me to the transplant room and introduces me to his nurse assistant, Galina Thaler. Thaler spent five years working with a hair transplant surgeon in Moscow, and another twenty here in the States. Chaffoo explains why I'm here, that I just need a few follicular units to bring to Stemson. She seems pleased. "The smallest case we've ever done."

She indicates a table with a padded ring, as for a massage. "You will lay down, you will put face over here." I do that. Thaler flips through sheaves of hair at the back of my head, location-scouting.

"You have very nice density."

"Thank you."

Chaffoo takes clippers to a small patch midway up the back of my head, a patch that will, some weeks later, make my hairstylist think I'd had brain surgery. He injects anesthetic, waits for it to take effect, and switches on the NeoGraft. This is a motorized version of the circular punch, activated by a foot pedal. Lying here, numbed and face down in a padded ring, all I experience of my follicular extractions is the sight of Chaffoo's foot pressing a pedal every few seconds and the accompanying burst of buzzy sound. He may as well be operating a sewing machine.

Later, on my way out, I will have an opportunity to peek in

on a more extensive harvest. It takes Chaffoo maybe a second to core out each follicular unit. Thaler follows in his wake, tweezering out the loosened, partway lifted cores and swiping them onto gauze in a Petri dish. The pair work hyperfast, which is impressive, because the angle of the coring blade must match the slant of the follicular unit, lest it be sliced in two. United Farm Workers sometimes posts videos of people harvesting crops—cutting asparagus or yanking parsley bunches, banding, trimming, setting aside, moving so surreally fast that you almost think the footage has been speeded up. This is like that. Also similar: the endless bending and hunching. Chaffoo had surgery for chronic back pain a few years ago. It's work you'd like to have a robot for, and Chaffoo has one: the Muskishly named ARTAS iXi. It stands, disinterested, on the side of the room. Chaffoo finds the technology isn't quite there yet.

"Okay! Done." Chaffoo wheels away on his surgeon's stool to power down the NeoGraft. My donor site is smeared with bacitracin, but no gauze is taped in place, because my hair would get stuck. On the flight home, I will feel the back of my head start to ooze. Do you fly Southwest? Don't sit in 11B.

Thaler begins packing up my follicles for the trip to Stemson. I remind Chaffoo about my request.

"Um, Galina? She wants to take one of the follicles, to put in her leg."

Thaler doesn't find this delightful. Relents anyway. Chaffoo numbs a spot on my left calf and scalpels a short, deep cut. Always strange to watch a blade sink into your flesh and feel nothing. Lips moving, sound off.

A follicular unit is selected—one with two hairs. Chaffoo pushes it onto the cut. It's not going in. Thaler *tsk*s. "The hair grow in *this* direction."

"I know, it's just—"

"Let me put it." A brief hockey brawl follows, my unit

ricocheting between sets of tweezers. Chaffoo sinks the goal. "Got it."

Antonella Pinto stands outside the lobby with a cooler. She is the associate scientific director at Stemson, here to collect me and my twelve uprooted follicular units and drive us over to the lab. Though she knows Chaffoo's staff well, she doesn't come in. The follicles have about fifteen minutes before they start to wither. No time for chitchat.

There is traffic. Pinto makes a face. "If you don't mind, I'm going to drive a little bit Italian-style." She is from Italy, got her PhD there. Her English is excellent, though I notice she sometimes uses the plural agreement for *hair*. "They are naturally wavy," she said at one point, speaking of her own hair in a way that made it seem like a community of unique individuals. I wondered if it was a grammatical slip, or the side effect of time spent studying shafts and follicles magnified to the size of garden weeds.

We make good time. Pinto parks, grabs the cooler from the back, and leads me, fast, across the parking lot and down the main Stemson corridor. White-coated staff are standing by to dissect my bits. They are after the building blocks: dermal papilla cells and keratinocytes.

After I leave today, my cells will be fed and kept warm, with the hope that they'll thrive and multiply. If and when there are enough of them, we'll see something magic begin to happen. The two colonies will be introduced to one another and will proceed to organize themselves, first in polarized camps, then migrating together and forming a primitive "organoid"— less than a follicle, more than a blob. The plan is to implant some organoids on a lab animal, so when I come back in a few months, I'll be able to see my own hair growing from the back of a mouse. Not because Stemson plans to sell Mary Roach hair for bald men to use, but simply to provide me with a rough demonstration of what can one day, it is hoped, be done to restore a bare scalp. (With actual patients, the follicles would come from their own cells—more on this coming right up.)

Pinto promises to send pictures as my cells multiply. And that's the last time we speak. After a brief email exchange and a very cool photograph of Mary 2.0 living and growing in a Petri dish in San Diego, emails go unanswered.

I email Stemson's CEO, who lets me know that Pinto has left the company. He puts me in touch with the new science director, and the two assure me I'll be welcomed back once they figure some things out. They're working on a new approach. Cellular me has been moved to a freezer for a cryo-nap.

Time passes. Eight months of it. Had the follicles on my calf prospered in their new location, the hairs would be four inches long. But they did not. Nothing came of it. Was that also true of Stemson Therapeutics?

It was not. Almost a year later, I'm invited back.

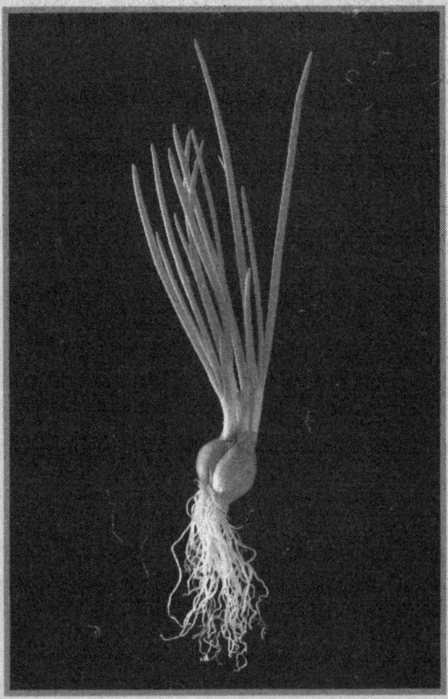

15

Splitting Hairs

Grow Yourself from Scratch!

The defense contractor General Atomics sits, as it has since the sixties, on a plot of well-tended green in San Diego's tech hub. Its buildings are arranged in rings—like the electrons of an atom, was the idea—around a "nucleus" containing conference rooms and a cafeteria. In among the lords of drones and lasers, twenty-eight men and women struggle to grow hair. Stemson Therapeutics has its offices and labs here. Early on I had imagined some connection between the follicle folks and their landlords—some shine-pated general on a personal crusade. It's not that.

It's the rent. "They charge half the going rate." I'm lunching in the nucleus with Kevin D'Amour, Stemson's chief scientific officer and something of a trailblazer in the field of cell engineering. Like any start-up, Stemson is in a race against diminishing seed money. D'Amour keeps a spreadsheet of the competition, which has thinned like a hairline, a half-dozen start-ups come and gone.

D'Amour, fifty-one, is athletically trim and good-looking, if not in the dashing romantic-lead manner the surname suggests. Before Stemson, he worked for another General Atomics renter, ViaCyte (now owned by Vertex), a firm closing in on a stem cell–based cure for type 1 diabetes. D'Amour's own hairline traverses the top of his head about where a set of headphones would rest, but the move to Stemson had little if anything to do with his extra forehead. "When I was looking for another job," he says, "my priorities were San Diego and stem cells." D'Amour has—his words—"stem cell fever."

I don't have fever, but I have caught the bug. This is not just the likely future of baldness. It's the likely future of medicine. Are some of your cells defective? Dying off? It's now possible to create new, healthy ones—your own—from scratch. The formal nomenclature for "scratch," in this context, is pluripotent stem cells. D'Amour has been explaining them to me between mouthfuls of black bean tostada, and I am now going to endeavor to explain them to you.

When you were young—unfathomably, microscopically young, an embryo of five, six, maybe eight days past conception—you were a blob of pluripotent stem cells: uniform cells bound for a multitude of bodily fates. You were an entity made of potential. To use the analogy of Mr. Potato Head, you were all potato.* Or plastic maybe. You get what I'm after. To stick to our flawed analogy, you can think of your DNA as the kid with the vision for the Potato Head she's about to build. Under its sway, your pluripotent stem cells began to differentiate. Some became blood cells, some bone, some neuron; humans have four hundred, even five hundred, cell types in all. The cells grew and multiplied and assembled themselves and soon there were

---

* The original Mr. Potato Head kit in fact had no potato. It was sold as a set of plastic body parts and accessories to be poked into an actual potato supplied by parents. After years of complaints—rotting potatoes in the playroom, children poking themselves and choking on tiny mustaches and pipes—Hasbro began including a plastic "potato body" and accessories large enough to pass choke-tube tests. The on-trend spud has continued to evolve. A Mrs. Potato Head arrived within a year. The pipe was decommissioned as part of a 1980s antismoking campaign. The current decade has seen a Mr. Potato Head Goes Green, made from plant-based plastic, a gender-neutral Potato Head set, and a low-carb Mr. Potato Head one-third the size of the original. Only one of those I made up.

recognizable pieces—eyes, teeth, mustache, hat: all the bits that together produced the delightful, one-of-a kind Potato Head you are today.

In 2006, Kyoto University researcher Shinya Yamanaka figured out a way to regress adult cells to their undifferentiated state. That is, he induced pluripotency, a state of existence normally reserved for the cells of very junior embryos. In theory, one can regress any cell type; for practical purposes, blood cells get used, because you can easily collect them without cutting someone open and because they produce reliable results.

The fact that induced pluripotent stem cells—or iPS cells, as Yamanaka's discovery is more succinctly known—can be created in a lab is quite miraculous. But it's only half the miracle. Researchers have figured out how to take iPS cells, these science-born pluripotent cells, and instruct them to become, in theory, whatever type of cell scientists wish for. You could literally grow new brain cells from snot. In the context of medicine, this suggests the possibility of creating whatever type of cells a patient lacks, or that need replacing. And because iPS cells can be created from the patient's own body, there's no need to suppress the immune system.

To clarify, we are not talking about growing whole organs in a lab. "That's a long way off," D'Amour says. We get up to bus our plates. Growing complex, multi-cell-type organs from pluripotent stem cells is, for now at least, "science fiction." In present iterations of the technology, it's small aggregates of immature cells that are produced. Clusters and patches. Depending on the disease, that may be enough to serve as a potent treatment, or even a cure. D'Amour uses the example of Parkinson's disease. Clinical trials are underway at another San Diego start-up, Aspen Neuroscience. Patients' blood cells are being regressed to pluripotency and then instructed to become the kind of dopamine-producing neurons that Parkinson's destroys. Elsewhere iPS-derived retinal cells, islet cells, and heart cells are being tested in clinical trials—putting medicine on the brink of being able to reverse or slow macular degeneration, diabetes, Parkinson's disease, and certain kinds of heart failure.

"And hair loss," says D'Amour. "Don't forget hair loss!"

At a conference in 2007, D'Amour heard Yamanaka tell the story of how he made his discovery—that is, how he figured out the process by which cells can be regressed to pluripotency. Yamanaka described it as "dumb science." You start with millions of cells. You swap out some genes, using a virus to get the editing machinery in there. Yamanaka began with fifty to a hundred edits. He watched to see if any cells regressed. If even one colony became pluripotent, he could isolate those and culture them—nurture them and grow them into pluripotent multitudes.

The next task was to narrow down the number of edits, or "factors," that it takes to achieve the regression. "Does it work with just these thirty genes? Yes!" D'Amour is saying. "And can you do it with just ten? Yes. It was like 'Name That Tune.'" Ultimately it came down to four key edits, now known as the Yamanaka factors. The process is still time-consuming, still expensive. But very exciting. In 2012, Yamanaka was awarded a Nobel Prize for his "dumb science."

For the purposes of its research, Stemson is using off-the-shelf pre-regressed iPS cells. The company's focus is on the other half of the miracle: perfecting the technique that will prompt iPS cells to become dermal papilla cells and keratinocytes, the building blocks of human hair follicles. I'm headed now to Tissue Culture Lab 2, where that work is underway.

When I first read about Stemson's work, I pictured row upon row of follicles and sprouted strands, a sort of trichological rice paddy. But agriculture is the wrong analogy. Stemson grows cells, yes, but more critically, it instructs them. Tissue Culture Lab 2 is less a hair farm than a hair kindergarten. Lisa

McDonnell, Stemson's senior manager of research and development, is showing me around. She bends down to open the door of an incubator. Inside are stacks of plastic flasks, each a sort of classroom for the cells within.

"So these are cells in the process of differentiation." McDonnell has round, friendly blue eyes, small feet, and a bald man's dream of hair, abundant and thick. Her default expression is an unforced smile. I've only known her ten minutes, but I think it would be difficult to dislike her.

She picks up one of the flasks. "These guys, today we're changing their media"—their pond water, if you will—"to the dermal papilla media." Having graduated from pluripotency to an interim stage called neural crest cells, the cells in the flask McDonnell is holding have just received a new set of instructions—new pond water, with different growth factors and nutrients: a proprietary blend of amino acids, lipids, and glucose. Immature cells take their cues not just from DNA but also from their environment: what they're fed and exposed to, signals from other types of cells, even how densely they're packed together.

"Right now we're telling them, 'Stop being neural crest cells. We don't want you to continue down this pathway. We want you to start being dermal papilla cells.' So they're getting some skin signals, some hair follicle signals. They're in there figuring it out, going, 'What do I do with this?' And—hopefully!—becoming dermal papilla cells." McDonnell says this with a genuine, almost breathless ebullience. You can scarcely imagine a more enthusiastic champion of regenerative medicine.

Established recipes exist for creating commonly studied varieties of medically relevant cells like neurons and liver cells and cells of the immune system. Only Stemson has worked out the formulas for turning iPS cells into dermal papilla cells and keratinocytes. It was a process of trial and error spanning months. There was tedium. Lots of hovering over huge grids of cells with loaded pipettes, air-dropping different mixtures over varying timelines, then waiting to see what turns out. "So much of science is waiting," McDonnell says. And when something finally works? "I cannot *describe* to you the satisfaction!"

The flask goes back into the incubator. McDonnell walks me through the rest of the lab. The donors' cells used for basic research are cryopreserved in three tanks, or dewars, of liquid nitrogen—minus 320 degrees Fahrenheit. Each tank holds eight shallow boxes in a vertical aluminum rack. The tanks stand together in a corner, high-rise housing for cells. People at Stemson call them R2-D2s, because they echo the shape and size (and thankfully not the vocalizations) of the shorter, squatter *Star Wars* robot. The size is typical for a lab like this, but it's possible to purchase much larger, room-sized cryopreservation units. In the future that D'Amour envisions, people will routinely have iPS cells created from their blood cells and then banked in massive cryo facilities—much the way people today bank sperm and eggs (both of which, by the way and good God, have been created from iPS cells).* Should the owner of the banked iPS cells go on to develop certain maladies later in life, their cells could then be thawed and coaxed to differentiate into the type needed. Banking would offer a potent advantage for treating, say, an aggressive cancer, because the time-consuming work of regressing the cells has already been completed.

I'd like to say hello to my cells, or at least see their apartment. McDonnell walks over to a computer to look up my donor

---

* So far only in mice. Human "in vitro gametogenesis," should it make its way into practice, presents intriguing scenarios. The ability to generate eggs or sperm from, say, one's blood cells, would present novel solutions for infertility. It would mean that gay and lesbian couples could produce offspring with genetic material from both partners. Also, in theory, a person so inclined—Elon Musk comes to mind—could use iPS cell technology to generate his own eggs, fertilize them in vitro with his sperm, and produce offspring with no genetic input from anyone else..

number, then wheels tank number 2 into the light. She pulls a thickly padded cryo-glove onto one hand, and the metal rack is lifted from the liquid like a french fry basket coming up from the oil. When it makes contact with room-temperature air, the liquid vaporizes and pours down the sides of the tank in a fog that pools at our feet. It's so *science*.* I press a fingertip to the metal frame, then jerk it back. Do not ever put your tongue on one of these.

"Let's see. Box 4 . . ." McDonnell slides out the box. Inside is a grid of small compartments. "Fourth row down, fifth row across." She lifts a vial. "Here you are." The label notes the type and number of cells within: 150,000 DP (dermal papilla cells). And there they petered out. To do any meaningful work with someone's cells, McDonnell says, it takes at least 450 *million*.

You reach a point in life where you figure you've catalogued your flaws. You're aware of all the ways your body and mind are below par, and you've made your peace with it. Then someone comes along and lets it be known that your follicle cells are weakly. (Also that they're strangely pointy. "They look a little like pig DP cells," said Antonella Pinto the first time I was here.)

McDonnell touches my arm. "Mary, not all donors expand well."

D'Amour, who had disappeared for a meeting, is back with us. He's going to walk me over to another lab. I've asked to see human hairs growing out of iPS-derived follicles. In actual skin, that is—and this has been the overriding bugaboo of Stemson's

---

* And at the same time, so heavy metal. Do rock stars get frostbite from dry ice? They do not, because dry ice (minus 109 degrees Fahrenheit) is nowhere near as cold as liquid nitrogen (minus 320°F). Scientists rarely get ice burns from liquid nitrogen, as they know to wear gloves. And, with the exception of a woman whose burns are detailed in the journal *Cureus*, they also know not to fish around in the liquid nitrogen with their bare hand when they drop one of those gloves into the tank. Happily, the woman made a full recovery from the burns, if not the embarrassment.

efforts. As with so much of regenerative biology, the challenge is not just the creation of cells and patches and pieces but also their survival and functionality within a living human being.

On the way, we detour for coffee. It's a quick walk through the inevitable San Diego sunshine. I stop to marvel at a Roomba-like mower making perfectly aligned passes back and forth across a lawn. D'Amour is minimally impressed. "Every now and then you see one stuck somewhere, spinning its wheels." It's a bit of a metaphor for Stemson's last couple of years.

When I visited a year ago, follicular units were being implanted into the skin of mice. There they produced hair cells, which was exciting, but the cells didn't necessarily grow in an upward, hair-like fashion. They grew sideways, downward, around in circles. They made "disorganized hair." It wasn't something investors were likely to be overjoyed with. *Look! We've grown small blotches of hair material under the skin of these mice!*

Something was needed to guide the growth—and to keep the skin above from closing up (that is, healing). Tissue engineers were brought in to design tiny biodegradable growing tubes. These were arranged in little racks, a setup Antonella Pinto likened to Barbie combs. But the tubes were too fragile to implant with tweezers. And if tweezers could crush them, however would a NeoGraft follicle transplanter or an ARTAS iXi robot manage it? The other problem was mouse skin itself, which is ten times thinner than human skin, thinner in fact than the length of the tubes.

This was all going down around the time I visited, when Pinto disappeared into the squid ink of an NDA. In the intervening year, Stemson completely rethought their approach. D'Amour presses a lid to his cappuccino. "When you have data that says, 'This just isn't working,' you need to have the courage to drop it."

The new solution is to let the iPS-derived follicle cells arrange themselves around a sort of wick, a short length of very thin suture thread. More tedium! The sutures have to be hand-poked though side-by-side droplets of cells, one droplet containing dermal papilla cells, the other, keratinocytes. The two cell types then organize themselves into a primitive follicle-like state, all the while clinging to the thread. The thread with its budding follicle is then implanted*—eventually, it's hoped, with a hair-transplanting tool like the NeoGraft—and then falls out, leaving the infant follicle behind, a couple of days later. Once implanted, it receives signals from the skin cells around it. It grows, it matures. Hair sprouts. The patient's confidence and self-esteem return, the sun comes out, the investors rejoice, and all is well with the world. Is the hope.

Nude mice are named for their pink, hairless appearance. You could be forgiven for assuming that they're mice engineered for hair loss research. In fact the nude mouse does have fur, albeit sparse and colorless, an old man's patchy white stubble. The appeal of the nude mouse, for medical researchers, is a

---

* Assuming one could anchor strands of thread, could one simply implant thousands of those on a balding person's head? One could, and one has. Beginning in the 1970s, some dermatology clinics began offering implants of "artificial hair" made from polyester or acrylic—basically wig hair—installed strand by strand into the scalps of men who lacked sufficient hair of their own for transplantation (and women with overall "diffuse" thinning). Using a sort of dermatological crochet hook, the "hairs" were knotted to connective tissue deep below the skin. As you might imagine, results were non-ideal. It was hair that could not be styled or dyed and would not go gray when you did. The fibers had to be spaced twice as far apart as natural hairs. Even a doll had more natural-looking coverage. The "fiber fallout rate" was between 10 and 30 percent per year, meaning one had to return regularly for re-crocheting—and for removal of the unsightly "sebum plugs" that tended to develop around the fibers. Infections occurred, and occasional serious complications (cardiac arrest, septicemia, hepatitis) prompted the FDA to ban synthetic hair implantation in 1983.

genetic mutation that leaves it unable to produce T cells and thus unable to mount an immune response to foreign tissue. So you can graft in place, say, a tumor you are studying and hoping to destroy, and the mouse's body won't reject it. The same holds true for the dime-sized patches of semi-solid human skin slurry that serve as the "soil" for the "seed" of Stemson's iPS-derived starter follicles.

How to make human skin slurry: pipette two types of skin cells into a gel of extracellular matrix. Place on mouse. Let sit. Over the course of eight weeks, the cells and the matrix will self-organize, like the *Terminator 2* villain who reassembles out of a silver puddle of himself. Into this "humanized" patch of mouse, the scientists of Stemson implant their nascent follicles-on-a-string.

A biologist named Alison Chambon, nearsighted and fighting off a cold, is fiddling with a microscope, preparing to examine one such patch. One can't tell by eye if the trials were successful, as there may be three kinds of hair going on: 1) wayward mouse fur; 2) the exciting, hoped-for hairs from implanted iPS-derived follicles; and 3) "background hair." Background hair is human hair that's grown from the skin slurry.

I have to interrupt Alison. The slurry grows hair? Could you instead just peel away a guy's bald patch and replace it with some of that? Skip all that expensive, time-consuming iPS cell regressing and differentiating stuff?

"We talked about that." D'Amour laughs. "A little invasive."

Alison pulls up a photograph on her laptop. It's a mouse that appears to have a patch of scrabbly white-guy chest hair. This unappealing thicket is "background hair."

Later D'Amour shows me an image taken with a camera-equipped microscope. In it, a single human follicle stands out, straight and long amid the stubbier mouse follicles.

"So it's working?"

"It's working," D'Amour says. His tone is a little anemic.

"How well remains to be seen." He estimates Stemson is two years* out from beginning clinical trials.

Here is something else that remains to be seen: Will a balding patient's brand-new iPS-derived follicles be sensitive to testosterone? That is, will they eventually wither and shrink like the follicles that led him to seek out Stemson Therapeutics in the first place? "We won't know until human clinical trials," D'Amour says, adding that the patient can always order up another batch. (Whereas traditional hair transplant patients are limited to a percentage of the follicles on the sides and back of their head.)

That sounds expensive, I say to D'Amour. We're back in his office, a modest, largely undecorated space that he shares with Stemson's chill and obliging CEO, Geoff Hamilton.† Earlier, D'Amour told me it cost Yamanaka $2 million to create the original iPS cells. After twelve years, he laughed, the cost is down to about $500,000. It is, as D'Amour put it, a bespoke process. You're talking about regressing and then differentiating each individual client's cells. In time, of course, the cost will drop. AI can be used to identify which cells are responding, and robotics can be developed to carry out many of the tasks.

A simpler, cheaper approach would be to use pre-regressed iPS cells—someone else's. But that would mean that the patient's immune system would reject them without some kind of immunosuppression regimen. And no one's going on immunosuppressive drugs just to restore their hairline.

Thus the Holy Grail of regenerative medicine is "stealth" cells—a line of iPS cells that are able to dodge the human immune system, making immunosuppression unnecessary.

---

* Start-up years are like dog years. You need a conversion factor. From my haphazard comparison of claims and actual market readiness, I'd say one start-up year equals three calendar years.

† No relation to dermatologist James B. Hamilton of the Norwood-Hamilton scale of male-pattern baldness. And currently a zero on the scale.

"That's where the entire industry is spending tens of billions," Hamilton says. As I write this, three companies are testing stealth iPS cells in cancer trials. Treatments for anything else are likely five or more years off. "Because God forbid you create an aberrant cell that can evade the immune system. This is the FDA's very legitimate concern."

Stealth cells may represent an exciting future for many patients, but they would be an imperfect option for hair loss. Follicles grown from universal cells will grow universal hairs. If those hairs are straight and brown, and the hair remaining on a person's scalp is, say, blonde and curly, that person will have a full but unusual head of hair.

Not long ago, walking through Target with his wife, D'Amour stopped to examine a jar of moisturizer. It was on sale, but that wasn't what caught his eye. It was the words WITH STEM CELLS. And there they were in the ingredient list: "apple stem cells." D'Amour points out that plant stem cells are found in the shoots and roots, not the fruit of the plant. While it's possible to culture them in a lab, that's unlikely to be the case in a cheap Target product. His guess is that the company just added some cells from apple stems.

"'Stem cell' has become a marketing term," D'Amour says. "People don't know what it means, but it's viewed as a good thing."

It's one thing to throw some ground-up fruit stems in a jar of face cream. Magnitudes skeevier is the large and wholly unregulated market for stem cell injections. You may recall from an earlier chapter that I spent some time in a hospital in a small, remote Mongolian city. Walking down a hallway one afternoon, I noticed a sign above a door that read Stem Cell Clinic. It was hard to imagine that a clinic in Murun, Mongolia, was regressing patients' blood cells to pluripotency. What exactly were they doing in there? What is it that they, and for that matter thousands of American orthopedists and other physicians, are injecting?

"Aspirate from liposuction," D'Amour answers. Goo sucked from your own belly or neck, cleaned up and centrifuged into different components. The part that gets used—the so-called stromal vascular fraction—is in theory a bunch of fat stem cells (or "adipose-derived stem cells"). It's not a defined stem cell population like what you'd obtain by growing out cells in a lab. But you can imagine the appeal, said Leah Bellas, founder and director of BellasFATLab (which does research, not stem cell injections), when we spoke by phone. "People are like, 'Wow, so I can lose fat and treat my osteoarthritis all in the same day?'" The clinics have sprung up everywhere. Bellas calls them Stembucks.

Does the treatment help? Unclear. "The studies are not robust or well executed," Bellas said. In part because the practice is unregulated. (If you're not editing or manipulating cells, if you're simply taking someone's own cells out and putting them back somewhere else, FDA approval isn't required.) Thus there are uncontrolled variables. What type of fat is it? Belly fat? Chin fat? How was it fractionated? Not only is the efficacy undocumented, so too are the long-term effects. It's a cellular wild west, and it's big money. "People want it, and they pay cash for it," said Andrew Rosenbaum, the orthopedic surgeon who teaches at Albany Medical College, and whom I also contacted by phone. "It's insane."

D'Amour concurs. "They're preying on people who are desperate. It's been a problem for over ten years, and it's getting worse." In 2017, three patients at a Florida clinic had stem cells from their liposuction glop injected into their eyes to treat macular degeneration. All had had some vision loss before the injections but were able to see well enough to pass a driving test. One went blind after the injections, and the other could barely see.

There is one thing fat injections are good at. These injections aren't going into joints or eyes but, rather, into buttocks and hips—in pursuit of nothing short of mathematical perfection.

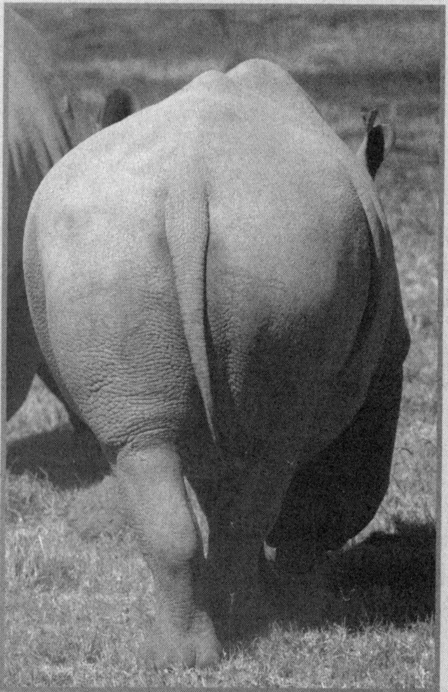

16

The Ass Men

Chasing Perfection with Math and Fat

In the history of gluteal enhancement, Mexico City stands out. It *protrudes*. It was here, in 1979, that a plastic surgeon, Mario González-Ulloa, first installed a pair of silicone implants designed specifically for the buttocks. The textbook *Body Sculpting with Silicone Implants* calls González-Ulloa the "grandfather of buttock augmentation." The early 2000s saw a new generation of Mexico City buttock transformation luminaries, notably Ramón Cuenca-Guerra. In his 2004 paper "What Makes Buttocks Beautiful?" Cuenca-Guerra laid out four "determinants of gluteal beauty" as well as the five types of "defects," with strategies for correcting each one. I, for instance, have defect type 5, the "senile buttock." (González-Ulloa's depiction of this took the form of charcoal nudes* contrasting "the typical 'happy buttock'"—high, rounded, dimpled—with its counterpart, the low-slung, drooping "sad buttock.")

While I understand the value of standardizing procedures and setting guidelines for surgical practice, I tripped over

---

* As late as 1991, González-Ulloa's papers continued to include art school–style illustrations. Often the illustrations showed where to insert the implants, but the nudes aren't depicted lying on an operating table. They stand as though posed in a figure drawing class, one arm crooked above the head or laid demurely across the breasts, oblivious to the disembodied arm reaching up from the margin of the paper to slip in a donut-sized silicone sack.

Cuenca-Guerra's methodology. How and by whom had the determinants been determined? Like this: 1,320 photographs of "nude women ages 20 to 35 years, as seen from behind" were presented to a panel of six plastic surgeons, who "pointed out which buttocks they considered attractive and harmonious, and features on which this attractiveness depended." Oho!

I thought it would be interesting to talk to Cuenca-Guerra about the notion of a visually ideal female figure.* As something that could or should be surgically created (or, in the case of the senile buttock, re-created). As something that even exists. I sent an email using the address on a more recent journal paper. There was no reply. Ramón Cuenca-Guerra's buttocks are in worse shape than mine. He has been dead for some time. I was able to reach a colleague of his, José Luis Daza-Flores. Here was the third generation; just as Cuenca-Guerra had studied under González-Ulloa, Daza-Flores had studied under Cuenca-Guerra, extending the lineage and making Daza-Flores, I guess, "the son of buttock augmentation."

Daza-Flores collaborated with Cuenca-Guerra on a paper called "Calf Implants," in which the team did for the lower leg what Cuenca-Guerra had done for the butt: laid out "the anatomical characteristics that make calves look attractive" and the "defects" to be addressed. Here again, plastic surgeons were recruited to judge images—2,600 of them, a vast photographic millipede of female legs.

The paper took an unexpected turn. Referring to a marked-up

---

* Some of the early work involved waist-to-hip ratios of *Playboy* centerfolds and Miss America contestants—in part because their measurements were publicly available. The winning ratio among these culturally anointed 1960s icons—0.7—was called into question in a 2016 paper by plastic surgeons at Loma Linda University. The team had obtained "the necessary licenses" to digitally alter an iStock photo of a woman's naked backside, creating eight different waist-to-hip ratios, ranging from 0.8 (boyish) to 0.5 (Kardashianish). The most "attractive" waist-to-hip ratio, as judged by 1,146 internet users, was 0.65. So, pretty much the same.

photograph of a lower leg deemed attractive, the authors tried to show that its measurements conformed to what is known in mathematics as the divine proportion (or golden ratio)—1.6 (I'm rounding it off) to 1. When you divide a line into two parts such that the whole length divided by the long part is equal to the long part divided by the short part, both those ratios will be 1.6 to 1. I found an illustration of the divine proportion on a website called Math Is Fun (and convincing no one). The golden dividing line splits the length such that one chunk is roughly two-thirds and the other is around one-third. The ancient Greeks divided the "ideal" face into similarly proportioned thirds. This was the first time I'd seen the divine proportion applied to a leg.

The paper contained sentences like this: "Seventeen women had thin legs, in the shape of a tube, and a mere 1:1.618 ratio in the A-P and L-L projections." Though I confess to not grasping the particulars of the discussion, I believe that to be a mathematically precise description of cankles.

The paper also cited the Fibonacci sequence, a progression of numbers wherein each successive number beginning with the second 1 is the sum of the previous two. It goes like this: 1, 1, 2, 3, 5, 8, 13, 21, 34, 55 . . . And, beginning with 5 and 3, if you divide two successive numbers instead of adding them—21 over 13, say, or 13 over 8—you will always get, ta-da, the divine proportion: 1.6.

The Fibonacci sequence turns up with some regularity in nature. You can find it in the turnings of the spiral traced by the inner chambers of the nautilus shell as it grows, or the growth path of sunflower seeds—two commonly cited examples. (Less commonly: the branching pattern of the sneezewort plant.) In humans, it's apparently not just calves. Calculator-happy physicians have published papers purporting to show the golden ratio in the relative lengths of the bones of a finger, the proportions of incisors in relation to canine teeth, the structures of aortic valves and the branching of the coronary arteries, the dimensions of the uterus at peak fertility, "optimal nipple position," and the helices of DNA.

It is an intriguing, if not entirely convincing, notion: that mathematics defines beauty.

When I wrote to Daza-Flores, he confirmed that when he transfers fat to the hips and buttocks, he follows the Fibonacci sequence. I pictured him with a purple surgical marker and a ruler, drawing arcs and equations on an anesthetized torso. How was he coping with what those in his field refer to as "the Kardashian movement": the surging demand for rear ends that fall well outside the Fibonacci ideal?

The plastic surgery clinic of José Luis Daza-Flores is on a residential side street in a pleasant Mexico City neighborhood called Noche Buena. The building also contains a day spa and a cosmetic dentist, but nothing about its exterior suggests the upscale beautifying that goes on inside. The waiting area seems designed to calm rather than to impress. The seating is comfy and the room is quiet, the clamor of a Mexico City weekday replaced by synthesized instrumental music and the occasional hushed spritz of a wall-mounted scent dispenser.

The lack of ostentation is in line with my impression of the surgeon himself. He is courteous, soft-spoken, deferential. This is a man who says *"por favor"* to Alexa when he wants to change the music in the operating room. When he speaks with patients in the waiting area or before surgery, I've seen him reach out to take their hand.

It's 8:15 a.m. now. Daza-Flores sits in the clinic's kitchenette, dressed in scrubs, having an espresso while the morning's patient is prepped for surgery down the hall. He looks young for his age, but not worked on. I don't see any of what his ilk like to call "defects."

From a folder, I slide out Cuenca-Guerra's "Beautiful Buttocks" paper. In the most respectful way possible, Daza-Flores distances himself from the work. Cuenca-Guerra, he says, concerned himself with the placement and size of buttock implants, but he overlooked the sides of the hips. It is this lateral projection that, coupled with a relatively smaller waist, creates the classic hourglass figure. You can make a pair of butt cheeks protrude from here to Puerto Vallarta, but it won't change the patient's figure viewed head-on. It won't make an hourglass out of a grandfather clock. In 1973, a few years before Dow Corning worked with González-Ulloa to design the first cosmetic buttock implants,* Tennessee surgeon William Cocke pushed a pair of silicone breast implants under the skin on the flanks of a woman who was "quite concerned about her underdeveloped hips." The results were described in one textbook as "less than optimal." Likely this was because implants placed superficially—beneath the skin and fat, rather than under or between muscles—tend to shift, and their contours are often visible through the skin. (In his writeup of the results, Cocke mentions none of this, noting simply that "the patient has since married." As though narrow hips had stood in the way of matrimonial bliss.)

"To reshape this lateral contour you need fat," Daza-Flores says. He gets up and takes a bowl from the refrigerator. The

---

* Dow Corning had also made the first silicone breast implant, circa 1961, this one at the urging of Texas plastic surgeon Thomas Cronin. Cronin had been "inspired by the look and feel of a bag of blood." It is hard for me not to picture the scene: Cronin standing around the OR, idly gazing at a bag hanging on a transfusion pole and turning to a colleague: *Hey. Does that remind you of something?* Cronin contacted Dow, and his resident, Frank Gerow, implanted some prototypes into dogs. The team were "excited" by the outcome—and again, stop me from imagining this. A few tweaks down the line, the silastic gel "breast prosthesis" made its debut—available in Small, Medium, and Large, as well as two sizes long since discontinued—Mini and Petite.

timing makes me think for a moment that he's retrieving a bowl of human fat. It's not, of course. It's papaya. He sets it on the table between us. "Fat transfer," he continues. That is, suctioning fat from one area and injecting it someplace else. The technique is known generally as liposculpture. Specifically and casually, when applied to the nether area (with or without implants), it's a Brazilian butt lift.

"So the surgery has changed," he says, "but the concept of divine proportions has remained."

I ask Daza-Flores if he can show me some examples of golden ratio divinity among his patients. He scrolls through photographs on a laptop. Lots of TikTok stars and beauty influencers. "These are pretty close to ideal proportions," he says. Here's the thing, though. Their befores look pretty ideal too. In 2021, a team of Norwegian and American plastic surgeons published a paper with the subtitle "Golden Proportions in Breast Aesthetics." Here's what stood out for me: Out of thirty-seven subjects, only five had breasts that conformed to the golden ratio. (An additional four had one golden breast.) "Neither breast was considered optimal for 28 (76% [of the]) subjects." The highest aesthetic ranking belonged to a "virtual subject"—a computer-generated torso. Plastic surgeons and their patients are chasing an ideal that rarely exists in undoctored human anatomy. The more work beauty influencers get done, either surgically or through TikTok filters, the shittier the rest of us feel about our utterly normal-looking bodies and faces.

And here's the other thing: by now, the desired look has eclipsed even Fibonacci's ideals and wandered into the realm of cartoon, of anime.

Before I left, I printed out a series of photographs from the London tabloid *The Sun*, purporting to show Kim Kardashian from behind in a thong bikini bottom. The woman's buttocks and hips dwarf the rest of her legs, a "defect" referred to in an *Aesthetic Plastic Surgery* paper as the "lollipop deformity," or "marshmallow on a stick."

"That to me is not beautiful," Daza-Flores allows. With his fork, he indicates the fold at the bottom of the woman's butt cheeks. "It's too long, and too creased." The intragluteal fold, as it is professionally known, should, he says, be more of a gentle ravine and extend no more than a third the width of the buttock. The folds of Alleged Kim's traverse the entire cheek. "It looks like a heavy buttock."

Is heavy. A set of the largest silicone buttock implants available weighs more than six pounds. A woman with double-D breast implants is hefting four pounds. When women complain of sagging breasts, Daza-Flores will suggest a lift rather than implants. "If you use an implant—for about six months the breast will look nice, but eventually it will fall down worse than before."

Daza-Flores slides the printout back to me. "She has everything you can get, all at once. Implants, fat transfer, injections of Sculptra," he says, referring to a filler. "She's probably getting something every six months. And girls ask for that. They say, 'I want the Kardashian surgery.'"

And who wins? Fibonacci or Kardashian?

"I try to suggest not going that far," Daza-Flores says. He tries to get inside their heads, to see why they're asking for this. Is it something their partner wants? He counsels patients against getting implants to please someone else. Because, as he puts it, the surgeries often outlast the relationships. He has had patients who've "changed out their breast implants every time they change boyfriends."

He also reminds patients that the look is a trend, and like all trends, it will fade. (Indeed, two years later, when I return to this chapter before going to print, the American Society of Plastic Surgeons is reporting an Ozempic-fueled trend for a leaner "ballet body.")

Daza-Flores drains his cup. "If I don't do it, they'll do it anyway. They'll go with another MD." Often, that MD is not a board-certified plastic surgeon. It may be an aesthetician or a general practitioner who got a certificate through an online course. Daza-Flores calls them "nomads," because once the lawsuits start to accumulate, they move to another state. One way to judge practitioners, he says, is to see how long they've been practicing in the same location.

Daza-Flores won't be using buttock implants of any size on this morning's patient, only fat transfer. Lately he's been using implants less often. The Brazilian plant that manufactures his preferred brand burned to the ground, and other companies have been having trouble obtaining or renewing their import licenses. Daza-Flores suspects corruption: government officials demanding exorbitant fees and bribes.

But the main reason is that José Luis Daza-Flores loves to work with fat.

Medical equipment catalogues fascinate me. It never seems to occur to the people who name the equipment and write the copy that anyone but a doctor will be browsing the offerings. Would you otherwise name your liposuction cannula The Fat Disruptor? The Bayonet Infiltrator? Clearing the path for the cannula is the "punch," which, in the words of the Black & Black catalogue writer, "make[s] the initial stab wound."

That's done now, and the cannula is in, nosing around the patient's lower back. The to-and-fro of liposuction looks very much like the motion used to wield that other wanded suction appliance. But quicker and more vigorous. More like mopping than vacuuming.

Daza-Flores waits for me to finish nattering. "Grating cheese. That's what it feels like," he says.

Because 20 percent of the suctioned fat is blood, the stuff moving through the tubing is dark pink rather than yellow. This, combined with the fat collection canister's resemblance to a blender, makes it appear that Daza-Flores is extracting raspberry smoothie from his patient. His wife, also an MD, had worried that I might faint or feel sick at the sight of fat being sucked from a body. If anything, it's making me hungry.

Once the raspberry treat begins to fill the canister, blood cells and serum settle out, leaving a layer of fat above. Daza-Flores uses the verb *decant,* which adds a touch of class to the proceedings. Some surgeons use a centrifuge to separate out the fat, but Daza-Flores feels that the forceful spinning can damage or destroy the cells. Even with the gentler treatment, around half the injected fat cells will die and be absorbed by the body.

The fat Daza-Flores loves most is the fat most of us hate. Love handles. The blub at the waistband. "This is the *fantastic* fat," he is saying. He loves it for the reason we hate it: it is stubbornly persistent. If a person diets, it's often the last fat to disappear. If they start to regain weight, that's where the fat first appears. In liposculpture, Daza-Flores moves the recalcitrant back blub cells away from where they're unwanted and installs them where they're wanted. "The genetic information remains," he says. Now if a patient puts on weight, it doesn't show up as back blub but as curvaceous hips and happy buttocks. Donor dominance!

For this same reason, Daza-Flores never uses back blub to fill out parts of the face. "If you take it from down here, and then they gain weight, the cheeks get fat."

Almost done. The cannula has widened the wounds made by the punch. They look like small slots, as though the woman has USB ports. An assistant switches out the cannulas, from aspiration to injection. In all, about six quarts of fat have been sucked from the patient, two of which will now be injected in her buttocks and the sides of the hips. This is where Fibonacci steps into the room. Daza-Flores starts at the outside flank, spirals down and around, and ends up at the center of the buttock. He likens the path he follows to the spiral of a snail shell, yet another example of Fibonacci numbers in nature. I had imagined him working from calculations, plotting points of the curve on the patient's skin. By now, he says, he can eyeball it.

The injection cannula has a plunger, which Daza-Flores squeezes as he guides the instrument backward. If you could see beneath the skin, it would look like someone decorating a cake. With human fat. *Happy birthday!* He pauses at the summit of the patient's left buttock.

"*Mira.* This is a dangerous area." Underneath the gluteal muscles are several large blood vessels. Even small amounts of fat in the circulatory system can spell trouble, in the form of a "uniformly fatal fat embolism." I'm quoting a 2018 Australian Society of Aesthetic Plastic Surgeons press release entitled "The Brazilian Butt Lift Named as the Most Dangerous Cosmetic Procedure." An embolism is a clot that breaks free and gets stuck in a narrow blood vessel, blocking the flow of blood. If it lodges in the heart or brain or lungs, that's serious trouble.

"There were a lot of deaths," Daza-Flores says. "A lot of girls dying from fat transfer. We didn't understand what was happening." Daza-Flores was part of a task force that reviewed the autopsy reports. In every case, fat was found in the muscle. Rather than keeping the injection cannula more or less parallel to the muscle, some practitioners were aiming down into it. Even if the angle was just 10 or 15 degrees off, Daza-Flores says, it was enough for the fat to enter the danger zone.

"They keep dying, to tell you the truth. Probably once a month here in Mexico City. Even now, people are still doing it, injecting into the muscle. There are clandestine clinics with poorly trained or untrained aestheticians." It happens in the United States as well. Women were dying in Miami so often that in 2019, the Florida Board of Medicine issued an emergency ruling. Florida Administrative Code R 64b8–9.009(2)(f) prohibits the injection of fat into or beneath the gluteal muscles.

I have to say, the patient looks great. Though in my estimation,* she looked great when she walked into the waiting room. I support everyone's right to do to their body whatever makes them feel good about themselves, but it would be lovely if feeling good about oneself didn't depend on altering the normal effects of aging. Particularly when the alterations involve the risks of major surgery. Interestingly, a majority of the surgeons in the aforementioned study of surgeons and patients and breasts seemed to feel the same. The surgeons, forty-five in all, were asked to imagine that a female partner or acquaintance had come to them expressing dissatisfaction with the size of their breasts. Would they recommend getting implants? Sixty-one percent said no.

Let's see about this one. "Dr. Daza-Flores, what would you say to your mom if she wanted you to do a full-torso liposculpture on her?"

He laughs. "I already operated on my mom." He extracts the cannula and hands off the empty syringe. "My mom loves surgeries!" He turns to the assistant. "Remember? We did the tummy tuck? But we didn't do fat transfer."

I recall what he said about embolisms. "Because of the risks?"

"Because she didn't want it. She doesn't need it!" An assistant hands him a new syringe of fat. "She's got a very big butt.

---

* Cosmetic surgery never needs to estimate. The profession abounds with rating scales. Are you over fifty? I urge you to steer clear of plastic surgery journals. You don't want to know how you score on the Neck Skin Laxity Scale or the Crow's Feet Grading Scale, or for that matter the Breast Aesthetic Scale, the Croma Dynamic Décolleté Wrinkles Assessment Scale, the Forearm Photoaging Scale, the Fitzpatrick Elastosis and Wrinkle Scale, the Forehead Lines Grading Scale, or the Midface Volume Deficit Scale. Trust me, you look fabulous.

"*Mira!* I'm making a fossette." *Fossette* is the medical word for dimple. A set of supragluteal (above the buttocks) fossettes is one of Cuenca-Guerra's four "determinants of gluteal beauty." Daza-Flores presses his forefinger into the woman's flesh to keep the fat from filling in that spot, first one side, then the other.

He steps back to assess his work. Removing the fat above the buttocks has given the patient a more defined waist and a longer-looking back. Bringing the torso closer to the golden ratio: two-thirds back and one-third ass. He makes a final sweep. The cannula moves around under the skin like a leg under a bedsheet. There are sounds, recognizable ones—sounds of sloppy eating, of things pulling up out of mud. I'm both disgusted and piteously envious.

One of the earliest aesthetic fat transfers took place in Heidelberg in 1895. The fat took the form of a benign tumor called a lipoma—essentially a fatty lump, and in this case apparently a sizable one. The lipoma was removed from a woman's back and deposited in a cavity in her breast where surgeons had earlier removed a more troublesome variety of tumor called an adenoma.

It must have worked reasonably well—well enough for the authors of the *Aesthetic & Plastic Surgery* article "History of Mammaplasty" to employ the word *heyday* in describing the embrace of dermal fat grafts that came later. These were not lipomas but rather chunks of everyday fat carved out of bellies and butts and stuffed into breasts. The main problem: "shrinkage." As much as 79 percent of these "free fat grafts" would disappear. The larger the chunk, observed surgeon John Watson in a 1959 *British Journal of Plastic Surgery* article, the more of it survived—graft shrinkage could drop as low as 45 percent. Watson reported better results when the dermis—the main layer of skin—was left attached to the fat. Skin, he reasoned, is "vaso-inductive," so perhaps new blood vessels grew in quickly enough to keep the transplanted fat cells alive.

And so it happened that for a brief period in the late 1950s, it was possible to visit a plastic surgeon to have twin wedges of buttock carved out, "dusted with penicillin," and transplanted in one curled slab, like a piece of rolled pork belly, skin intact. A half-pound serving per breast.

Watson's article includes a page of step-by-step photographs outlining his technique. Here you begin to see why the heyday was short-lived. Figure 2A is a photograph of a woman's rear with a gaping (3 inches wide) diagonal blood-red gash across one buttock. Picture the open red mouth of Bert, of *Sesame Street*'s Bert and Ernie. Figure 2B shows the aftermath: 9 inches of Frankensteiny stitches reinforced with a pair of dainty buttons—buttons!—one on each side, linked with a winding of surgical thread. Plus a notably deflated tush.

You can imagine how a few injections of filler presented a more appealing option: less pain and downtime, no scars, no sitting on sewing notions. The filler would need to be thick enough to pass as breast tissue yet thin enough to pass through the opening of a syringe. Before the arrival, in the early 1980s, of the liposuction aspirator, the substances injected were not typically fat. They were, it truly seemed, whatever fat-ish substance some enterprising plastic surgeon's gaze happened to land on. Some took their inspiration in the kitchen (olive oil, vegetable oil), some in the barnyard (goat's milk, cow collagen, pig collagen) or the forest (beeswax, tree resin derivatives), others in the supply rooms of industry (paraffin, petroleum jelly, various glues and polymers).

In one fateful case, inspiration arrived on the docks of Yokohama harbor during World War II. American army quartermasters, reports M. Sharon Webb in "Cleopatra's Needle: The History and Legacy of Silicone Injections," noticed that vats of transformer insulating fluid were vanishing shortly after they were unloaded. The fluid was essentially silicone. Japanese cosmetologists were, she writes, injecting it "into the breasts of Asian prostitutes who sought a more Western appearance to cater to the American servicemen." The story aligned with things Webb had heard from her father, who served as a medical officer in Japan during the war and knew an army master sergeant who'd made some side cash by giving the injections himself.

Thus began the practice that launched a thousand stripper careers, a few deaths, and as many class-action lawsuits. Free-floating silicone tended to provoke inflammation and other systemic responses, as well as infections, painful cysts, scar tissue, and droplets that would break free and migrate elsewhere.

It's hard to imagine the ignorance—or maybe just callousness—that would lead someone to think it would be possible to persuade the human body to accept, without complaint or consequence, a cup or two of hastily injected transformer fluid. Anyone with a solid sense of the complexities of the human form would also, you'd think, have a sense of the risks and challenges of trying out a substitute. It's just very, very hard to compete with the human body. That's why we still recycle it.

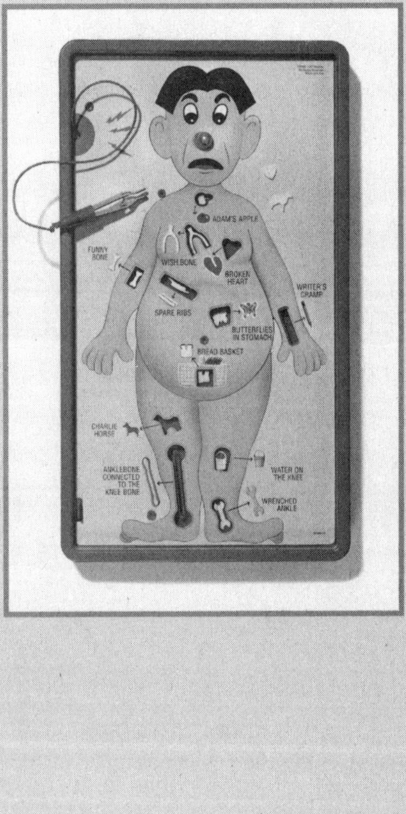
ADAM'S APPLE
FUNNY BONE
WISH BONE
BROKEN HEART
WRITER'S CRAMP
SPARE RIBS
BUTTERFLIES IN STOMACH
BREAD BASKET
CHARLIE HORSE
ANKLEBONE CONNECTED TO THE KNEE BONE
WATER ON THE KNEE
WRENCHED ANKLE

17

Some of the Parts

A Day in the Life of a Tissue Donor

If you're inclined at all toward donating your remains, I can tell you Pittsburgh is an excellent place to die. Your abandoned hull will make its way to the Center for Organ Recovery & Education (CORE), an organ procurement organization in a hilly business park overlooking the Allegheny River. Along with the titular organs, CORE parcels out the lesser pieces—known collectively as "tissue"—to the major suppliers of hospitals and surgical centers. Professionalism is a point of pride. The firm's Baldrige Award, a presidential distinction, rotates on a small dais in the lobby. The grounds are manicured to the point of topiary. The company dress code prohibits stubble. Even the building itself has a dress code; window shades must be lowered to the same level. Because, in the words of CORE's CEO, "Everything speaks."

If your business is body parts, you are speaking into a powerful wind of suspicion, urban myth, and bad press. CORE is a Tier 1 organ procurement organization, a rating achieved, in 2023, by just 26 percent of organ procurement organizations (OPOs). Almost half were Tier 3, the lowest rating. Causes for negative publicity are spelled out in a 2023 U.S. Department of Health and Human Services audit report: fraud, lack of oversight, misspent taxpayer dollars. Needless to say, OPOs are wary—warier than usual—of allowing outsiders in. But if the complaint is that journalists only pay attention when things are done wrong, why not let one in to see things done right? In terms of public opinion, part of the problem, I argued in seeking to gain access to a tissue recovery, is that the process has been so assiduously hidden. If you leave the postmortem extraction of bone, fascia, and tendon to people's imaginations, the scenes they imagine are likely more disturbing than the reality.

It took some time to locate someone who agreed with me. TJ Roser is CORE's funeral director/coroner liaison, the brave man who has offered to have me spend a couple of days on-site. We're sitting in his cubicle on the main floor of CORE, upstairs from the organ and tissue recovery suite. TJ's post exists because funeral directors, for different reasons, may also balk at removing a loved one's inmost structures. They'll try to talk families out of it, he says, because it makes their job harder. Embalming fluid can no longer be pumped through a single artery. And because of the extra time spent in refrigeration, the body starts to dry out and the blood may coagulate. Mortuary blood thinners will have to be administered to prevent clots from damming up the embalming fluid and causing limbs to swell unpresentably.

TJ volunteers that he had a blood clot in his leg a couple years back. He pulls up his pants legs to show me his compression socks. TJ Roser defies stereotyping. He is no type. He's a young, fit guy wearing compression socks. He's a mortuary professional who'll happily volunteer the (rather plump) markup on a direct cremation. He's a delight.

Despite advances like ceramic bone fillers and 3D printed metal implants, demand for donated human tissue is high. Donors aren't scarce, but *eligible* donors are. Ninety-six percent of people who consent to donating tissue are ruled out because some element of their medical history or social behavior has created an unacceptable risk for the people who would receive their tissue. Did you have metastatic cancer? Alzheimer's? Ever use IV drugs? Spend time in jail lately? You, dead person, are out of the running. Yale has a higher acceptance rate.

In a few minutes, I'll be listening in on a risk-assessment interview with the family of a potential donor, a retired (now extremely so) coal miner. A tissue coordinator will be speaking by phone with the man's widow and daughter. It is a critical step. Because regardless of a person's wishes or the dot on their driver's license, tissue can't be recovered until the next of kin has answered the eligibility questions. If they don't want their loved one's tissue donated, all they have to do is hang up.

TJ leads me down a hallway to visit the tissue coordinator. Digital donor tracker boards are mounted on the wall at intervals, like airport arrival and departure boards. These display the names of people who've recently died, or are expected to soon, in area hospitals—people under consideration as organ and tissue donors. Today's departures include an on-the-job electrocution, a gunshot, a drug overdose, and two fatal allergic reactions: a beesting and a suicide by peanut butter. (The deceased had a nut allergy.) The time of death is also noted, and it is a key piece of information. Once a hospital notifies CORE of a patient's death, the team has twenty-four hours to secure the medical examiner's release, have the risk-assessment conversation, get consent from the family, and begin the tissue recovery. If the body "times out," that's that. It's a wonder anyone's parts make their way into anyone else.

The coal miner's widow has concerns about the viewing. Her daughter is speaking to the coordinator, who has speakerphone on, and the widow can be heard at times in the background. The coordinator assures them that the mourners won't be able to tell. "The tissue is taken from the back and the lower parts of the body," she is saying. *Tissue* is a medically accurate word, but its other associations perhaps lighten the impact. Ligaments and fascia as tidy and easily accessed as a Kleenex.

"And his eyes will be closed."

"You mean they take the eyeball?" Some alarm registering.

"Just a clear film over the colored part of the eye." She's referring to the cornea. Medical terms are avoided during these calls and careful attention is paid to language. Tissue is "recovered," never "harvested," as the latter might seem to equate a loved one's remains with agricultural products.

"Tell me again," the daughter says, "what would they be able to take?" The coordinator puts each type of tissue in the context of the person it will go on to help. "He'll be able to give the gift of eyesight to two different people. Skin can help with skin grafting for burn victims. Bones and tendons from the legs can be used for reconstructive surgery. For cancer patients. Or for athletes that may need an Achilles heel." She meant to say "tendon," though heels are used too.

The widow gives consent, and the coordinator turns to a list of questions about medical conditions the miner might have had. These range from the mundane (pink eye, varicose veins) to the exotic (Ebola, Zika virus, mad cow disease). Not all are automatic disqualifiers. The answers and the relative risks they pose are considered together in deciding whether to accept or decline the body and which parts to recover.

Next come questions about the coal miner's social behaviors: prison time, sexual practices, drug use, and (because of the needles) body piercings and tattoos. Though it is unlikely that an eightysomething retired coal miner got permanent lip liner over the course of the past year, protocol requires asking.

The coordinator slides her pencil down the sheet. "The next few questions are a little more personal. I apologize in advance," she says. She begins with STDs. No, no, and no.

"Was he sexually active in the past five years?"

"Hey, Ma, in the past five years . . ." The daughter covers the mouthpiece. There's some muffled back-and-forth. The widow's tone shifts from wary to annoyed. The daughter persists. She is a nurse and a supporter of organ and tissue donation. She returns to the line. "That's a no." The *no* makes things easier for the coordinator, because now she doesn't have to ask about sex with other men.

The coal miner had some chronic lung issues, but nothing in the family's answers makes him ineligible to donate bone, skin, fascia, tendons, cartilage, or ligaments. He will shortly become tissue donor number 41 for the month. ETA 11:00 p.m., a job for the late-night shift.

I am no stranger to things that go on in cadaver labs, but what I am witnessing now I did not expect and cannot imagine having to deal with. A recovery coordinator named Lindsay is bent over a sheet of peel-and-stick labels with UPC barcodes, on which she must handwrite, up to ninety times, the donor's ID number. Because the fallout from a mix-up could be serious. After the tissue recovery is completed, hours from now, the packaging will begin. More labels, and scanning of labels, and entering of codes for two-step verification. As much time is spent on documentation and shipping of a donor's tissues as on their removal. You're expecting *The Jeffrey Dahmer Story* but it's closer to UPS.

Lindsay cracks her neck. "This is the worst part of the job." It's a statement that will, over the course of the next four hours, become a little bit surprising.

Lindsay's coworker introduces himself. He gives his first name as Don, but Lindsay calls him Donny G., which to my mind fits him better. The coal miner is here, too, a mounded presence on a gurney, mute but impossible to ignore.

Donny G. walks to a sink in the corner to scrub in. "I'll do the corneas."

"Knock yourself out." Lindsay still bent to her labels. She's less talkative than others I've met at CORE, and it's hard to get a read. The stark eyeliner and dark nail polish send you off in one direction, but her scrub cap—pink with butterflies and flowers—yanks you back. Lindsay's previous job had her in hospital operating rooms assisting surgeons. She prefers the current job. "They were shitty," she says of the surgeons. For the dead, she uses the adjectives of children's bedtime books: *fluffy*, not *fat*; *furry* for *hairy*. Respect is in the drinking water at CORE. They talk about it on the website. It's the name of one of their conference rooms. It's the reason protocol asks for a moment of silence, gurney-side, before a recovery begins.

Donny G. stands at the donor's head with a small bottle of Betadine. This is used to sterilize the surface of the eyes. Five brown drops carefully counted. I picture my mother adding vanilla extract to cake batter. *Get out of here, Mom.* I've always imagined the human cornea as a contact lens, but it's thinner than that. Using a scalpel, Donny G. first scores the cornea's perimeter, then switches to surgical scissors. Finally and slowly, with forceps, he peels the cornea away and drops it in a small jar of preservative. Further down the supply chain, someone will mark the cornea with an *S*, so surgeons can easily check that they haven't installed it inside out.

"Lyin' Eyes" has been playing as Donny G. does all this. As the night goes on, I will hear "Wanted Dead or Alive," "Another One Bites the Dust," "Spirit in the Sky," and "Only the Good Die Young" and will come to believe there's a Spotify playlist for tissue recovery.

With the "corns" done, the coal miner is wheeled to the prep room to be cleaned and disinfected, all over, even—or especially—the belly button. ("They're filthy," someone told me earlier.)*

---

* In the extreme, the filthy human navel will occasionally harbor an omphalolith—a navel stone. These are dark, rounded concretions of sebum and skin cells that have sat in the belly button so long they've practically fossilized. Omphaloliths are rare, typically turning up in people with deep belly buttons and shallow attention to hygiene—and, of course, dermatology journals. You can find case reports with lots of close-up photographs in the article listed in the sources, but I cannot recommend this activity.

For the first half hour, tissue recovery looks like a spa treatment. Lindsay and Donny G. share the work, gently sponging and then shaving the donor's exterior. He's a little fluffy, so the process takes some time. The gurney is tilted for a rinse. It's strange to see a human form that does not flinch or make a sound when cold water hits. (The dead do get goose bumps, though not from cold. As it does with the body's other muscles, rigor mortis acts on the hair follicles' arrector pili.)

It's time to turn him and clean the other side. Donny G. does the majority of the heave. As gravity assists with the last few inches, an arm swings over and smacks Lindsay's cheek. "He bitch-slapped me! You made him bitch-slap me!" She's stunned but also laughing. We all are. At her, not him. The look on her face.

"I'm sorry," Donny G. manages to get out, then dissolves into laughter again.

The donor is on his belly now, though not lying completely flat. "Because of his pannus," Donny G. says. As I have never heard this word and as Donny G. hasn't been using a lot of medical terminology, I mishear and register some surprise. Donny G. explains that "pannus"—or "pannus stomach"—is the technical term for an overhanging belly, aka "apron belly."

More soaping and swabbing. Lindsay wonders aloud whether it was the first time the coal miner ever slapped a woman.

Donny G. imagines so. "He seems a gentle soul."

The donor is wheeled back to the recovery suite, where he gets a final round of disinfection, this time with rubbing alcohol. We sit a moment, letting the alcohol do its work. Something about this guy, the way he's positioned on the gurney, the hairless pale topography of him, calls to mind a Francis Bacon painting.

Donny G. gets ready to recover two wide strips of skin from the back. He places rectangular paper templates on either side of the spine to guide the scalpel. Mom is back, her sewing patterns pinned to a spread of fabric. Skin comes away easily along anatomy's natural strata. Donny G. folds the pieces as best he can and slides them into preservative inside a heavy-duty plastic bag, which he sets aside. The areas without skin are red with some yellow patches where fat has remained. It doesn't look horrifying to me. It looks like the top of a lasagna. The donor is about to be turned over again, so however it looks, it's quickly gone from sight. The two get into position for the maneuver.

"There's gonna be purge," says Lindsay. "I can handle a lot of things. Purge I cannot handle."

Donny G. expounds: "'Coffee ground purge' is material coming up from the stomach. 'Milky' is—"

Lindsay exhales. "Alright. That's enough."

There is no purge.

Lindsay steps back and adjusts her face mask. "Here comes the fun part." "MS," they are calling it, short for musculoskeletal. The particular tissue processor who will receive the man's donations takes the following: bones (from the thigh, calf, knee, and heel), five different tendons, the patellar ligament, and the meniscus. They've requested the Julia Childesque "leg en bloc," meaning the whole intact interior of each leg; they themselves will dissect out the particulars. Thus Donny G. and Lindsay will be engaged mainly in freeing the inside of each leg from its moist and sundry moorings.

Donny G. applies a nine-by-eleven-inch adhesive-backed sheet that he uses to pull the pannus upward, away from the hips, where they'll eventually be working. He steps around to the right leg, Lindsay taking the left. With a scalpel, the skin is slit from the hip to the top of the foot, then coaxed to either side, with a kind of peeling and pulling motion. My notes say "like opening a corn husk." But that's kind of silly. We're not eating tamales here. We're on steps 3 to 4 of the International Hunter Education Association guide to field-dressing a deer. Just doing what has to be done to gain access. It is done cleanly and with precision.

The en bloc removal is complex, at times aerobic. More Francis Bacon paintings. Donny G. narrates flatly, pointing out the anatomical landmarks that guide the process. Like Lindsay, he's a little hard for me to read. As a kid, I used to go on trail rides at a stable near our house. I could never get a sense of how the horse felt about it. Did it mind carrying me around? It didn't seem to, but it also didn't seem to enjoy it. That's the vibe here. I'm another thing they've been asked to do. *Disarticulate hip, remove fascia lata, endure Mary Roach.*

At last the bloc is freed and lifted away, the leg stepping out of its own skin. Donny G. holds the bent limb under one arm like a bagpiper. With the other hand he pulls a plastic bag over it. The next phase is his favorite: restoration. Over the course of the next fifteen minutes, the anatomical chaos on the gurney will go back to being legs and hips.

He sets within the skin a pair of biodegradable, "cremation-friendly" (non-melting) bone replacement prostheses. (Restoring the rigidity of the legs makes it easier to carry and move the deceased.) These rods are lengthened or shortened to fit, in the manner of a vacuum cleaner wand. Standing in for missing muscle and kneecap are wadded sterile blue surgical cloths. All that remains is to pull the skin back up to where it was and truss the incision. Donny G. executes a perfectly even double baseball stitch, something he was taught in mortuary school. It's impressive, though since this man's family will be having a viewing, the stitching will likely be taken out and redone at the funeral home, so staff can get in and apply a topical embalming solution.

Three a.m. now. Except for the seams, the coal miner looks as he looked when he got here. Until you see it, you can't really imagine how it's possible. It's like trying to imagine a caterpillar reassembling inside a cocoon. Awe is the overriding feeling—for me, anyway. "That's why we like people to see the whole process," Donny G. says.

Lindsay nods her agreement. "If someone just walked in in the middle, they'd be like: 'Oh, *hell* to the no.'"

Having spent time in operating rooms, I can tell you that statement is true as well for major surgery on the living. And this operation, unlike any upon the living, requires no pain meds, no recovery time, no stint in rehab. I hope that what I've relayed here will encourage, rather than discourage, you to consider donating. As with any surgery, it's what comes of it that matters. As many as seventy-five people can benefit from one person's donated tissues. Earlier, TJ Roser shared a folder of letters from grateful patients. "I'm 21 years old," begins a note from a motorcyclist named Greg. "I was recently hit by a person turning into oncoming traffic. My injuries required an extensive reconstruction surgery. . . . I don't know if my gratitude can help your healing, but I want you to know you and 'your loved one' have made my life better."

Four a.m. I leave Donny G. and Lindsay to finish up in the shipment room, double-bagging and icing, labeling and boxing. The lobby is dark. The Baldrige Award has ceased its pirouettes. In an hour, a courier van will pull into the loading dock, and the coal miner will be off on his various ways.

Last Thoughts

I wasn't too far along in the research for this book when I began to suspect what I now believe: that even the simplest part of the human body defies efforts to re-create it. This led me to ponder what the simplest part actually was, and whether perhaps an equivalent artificial version had in fact been perfected. Tears, perhaps. Saltwater with a layer of oil to keep it from evaporating, right? How hard could that be?

Poking around for answers, I came upon what appeared to be a laboratory devoted entirely to the study of human tears. Naturally I wanted to visit. I imagined ads in a campus newspaper ("Get paid to cry"), people showing up to watch sad movies with little test tubes affixed to the corners of their eyes. There would be a walk-in filled with a thousand vials of frozen tears. Good chance of a hulking bronze eyeball sculpture out front.

I never went to visit, because there is no lab. TearLab turns out to be a tabletop testing device for diagnosing dry eye disease. It was developed by tear researcher Benjamin Sullivan. For the past decade, Sullivan and his company Lubris BioPharma have been working on a replacement for the human tear film. The tear film is distinct from "emotional tears," the high-volume saltwater that courses down your cheeks when you cry or something irritates your eyes. Sullivan called it an exquisite structure, multilayered yet thin as Saran Wrap. Emotional tears, as opposed to the tear film, would in fact be easy to replace, but there's no reason to do so. (Sullivan's father, the founder of the Tear Film & Ocular Surface Society, spent some time studying emotional tears. "I remember him collecting my tears when I was five years old," he told me. "I'd be crying, and he'd go, 'Wait, stay right there!' and come up to me with" —*yes!*—"a little test tube.")

Sullivan and I talked about the tear film for quite some time. We reached the forty-five-minute free Zoom limit twice. I can't begin to do justice to the complexities of this overlooked, essential, taken-for-granted eyeball moistness. Sullivan effused about a structure I had never heard of: the glycocalyx. This is a brush-like scaffold that extends into the tear film and keeps the various layers stable, as well as helping prevent evaporation by glomming onto water. The tear film contains mucins as well, gummy webs of long-chain sugars that trap debris and bacteria and deposit them as harmless gunk in the corners of your eyes. "The garbage trucks of the tear film," Sullivan said.

Sullivan's passion is lubricin, a friction-reducing substance cranked out by cells on the surface of the eyes and the inside of the eyelids. Without lubricin, the blinking eyelid drags and tugs rather than glides, and over time this can damage the cornea. Sullivan sent me a paper in which he and Lubris cofounder Tannin Schmidt figured this out by installing a cadaver eyelid and a cadaver cornea on either side of a lubricant bath. The paper included a photo of the inside-out cadaver eyelid, still with its eyelashes. Did you, like me, know a kid in grade school who tried to impress the girls by flipping his eyelids inside out? It looks like that.

Tears—the kind you cry—are wet, but they make your eyes feel dry. The salt load stresses and damages cells on the surface of the eyes. When these cells are stressed, they stop maintaining the glycocalyx, because it's an energy-expensive undertaking and they're trying to conserve resources to stay alive. The glycocalyx collapses, and lubrication ebbs. This is one reason your eyes burn after you cry. Another one: Tears wash away a lot of the good stuff, like lubricin, on the eye's surface. The same is true of commercial eye drops and "artificial tears." Thus overusing them, Sullivan said, can make dry eye worse. Drops don't come close to being a replacement for the tear film. "Some of the newer ones work for an hour maybe," Sullivan said.

Over the past decade, Sullivan developed a new kind of eye drop, with lubricin. It worked so well that a large pharmaceutical company bought the licensing rights. They ran a clinical trial, but they hadn't followed Sullivan's patented production process, and their own, less costly process destroyed the lubricin. The drops didn't work. More test tubes of Benjamin Sullivan's tears. Even when you nail the science, you have to grapple with the realities of business and profit.

So indeed, there is as yet no lab-made replacement for the human tear film.

What there are, Sullivan said, are autologous serum eye drops—drops made with a diluted component of one's own blood. In a 2020 review paper, autologous serum eye drops outperformed artificial tears in seven different trials. In other words, the best thing you can use is some other magnificently complex, multitalented bodily substance that evolved over millions of years.

I looked into tooth enamel, too. Had anyone succeeded in regrowing the shell of a worn or decay-pitted tooth? That is, with something equally strong and long-lasting? I tracked down a professor of biomedical sciences and bioengineering at the University of Southern California, Janet Moradian-Oldak, who is working on it. Tooth enamel, the body's hardest tissue, is made of biominerals—crystals of atoms of calcium and phosphate arranged in a highly organized manner. "Order creates stiffness," Moradian-Oldak said when we spoke. It had the ring of a universal truth. Hard, stiff materials tend to crack or shatter on impact, but less so tooth enamel, because in between the highly organized nanorods of the crystals is a highly unorganized arrangement of atoms and molecules, a material that provides cushioning and flexibility. "How the cells and proteins control all of this has been the passion of my career for the last thirty years," Moradian-Oldak said.

Had she succeeded? Was tooth enamel in fact replicable and replaceable? Moradian-Oldak replied that while she had recruited proteins that could successfully control the orientation of the tiny rods of crystalline material, the thickness she was able to create was limited to tens of microns—about as thick as a grain of pollen. "Not enough for dentists to work with." And though it was similar to natural enamel, it was not the same.

In the meantime, researchers elsewhere have created a synthetic tooth enamel even harder than the real thing, yet with some of the same impressive flexibility. Alas, there's no known way to incorporate it into a tooth. And the process used to create it involves extreme temperatures and materials not easily procured. "How practical will it be in a dental clinic?" said Moradian-Oldak.

Nonetheless, she considered the work a success, because an exceptionally hard yet flexible enamel has valuable applications outside of a mouth. She gave the examples of better helmets for soldiers or building materials that could withstand the shear forces of a powerful earthquake. In other words, rather than a new development in materials science inspiring a way to rebuild the body, the body had inspired a new development in materials science.

Fifty years from now, I have no doubt, all the glittering promise in all the chapters here, the materials and devices and techniques, will have been realized. The benefits will have been teased apart from the risks, the processing regulated and scaled up, the costs lowered enough for insurers to cover. Expect miracles. Just don't expect them overnight.

And don't expect too much of them. You can replace many of your original parts, and the replacements will improve in time, as they have already, in astonishing ways. But the pieces science builds—and grows and prints and scavenges—won't ever quite equal the ones we start out with. That's not a complaint, or a condemnation or belittlement of the progress that's been made. It's just me saying something we all know but regularly lose sight of. The body's all-day, everyday achievements—the architectural brilliance of cartilage or tooth enamel, the effortless autofocus of the eye, a heartbeat so committed it persists outside a body—these are the real miracles. That's where the gee-whiz belongs.

Epilogue

As this book was heading into production, in February 2025, the Trump administration announced drastic cuts in NIH grant money. Several of the researchers I had visited or spoken with depend on NIH funding. Muhammad Mohiuddin, director of the Program in Cardiac Xenotransplantation at the University of Maryland School of Medicine, has one major grant under review and two others submitted, all now imperiled. "We definitely need NIH support to continue our work," he said by email. The new federal health research funding agency ARPA-H, whose PRINT program would have helped fund Adam Feinberg's liver-bioprinting project, could shut down. Even grants previously awarded are under threat. Feinberg's colleague Jaci Bliley pointed out that many of these grants have signed contracts; thus no one is quite sure how things will play out. The University of Michigan Extracorporeal Life Support Lab, too, operates on a multiyear federal grant. "I don't need to tell you," said Wyeth Alexander, "how detrimental these cuts would be, both to the field and to those it serves."

In January 2025, I heard from Geoff Hamilton that, owing to an investor-funding shortfall, Stemson Therapeutics had closed its doors. No other biotech company is pursuing a stem cell–based cure for baldness.

Some encouraging news on the xenotransplantation front: a

patient at NYU Langone Transplant Institute who received a gene-altered pig kidney in November 2024 has broken through the two-month xenotransplant survival ceiling. (The organ failed at four months and was removed.) A clinical trial, with six patients to receive a United Therapeutics UKidney, is set to begin midyear 2025. In March 2025, surgeons at Xijing Hospital in Xi'an, China, announced they had transplanted a gene-edited pig liver into a beating-heart cadaver—a first—where it functioned with no signs of rejection for the duration of the study (about ten hours). However, the original liver was left in place, so it's hard to say how well the porcine organ performed.

The charity Exovent is finishing up the electronics for the Exovent negative-pressure breathing unit, and it is looking streamlined and spiffy. They hope to begin clinical trials later in 2025.

Sadly, the biofilm researcher Paul Stoodley died in April 2024. He was sixty-three.

Mark Randolph has met someone new (and cleaned up the tool room).

ACKNOWLEDGMENTS

Behind each chapter in this book is someone who said yes. Yes to a writer they'd never met, probably didn't have time for, had no reason to trust. What a gift that is. Especially since there is also, behind each chapter, someone—or two or three or five someones—who said no. My weightiest thanks, then, are for the generous, accomplished, brave, and just generally outstanding yes-sayers, the people who, no overstatement, made this book possible. In chapter order:

Jeremy Goverman (*yes, you may follow me around like a spaniel for two days and ask questions my institution would prefer I ignore*); Yi Wang (*yes, I'll take you on a road trip I don't have time for, set up interviews and translate the answers for you, take you out for hot pot, and vouch for you when the Chengdu immigration people suspect you may be a spy*); Bob Bartlett (*yes, you may visit my lab even though the public affairs people said no because they think you're secretly an animal rights fanatic*); Maurice Garcia (*yes, I'll spend an evening answering your oddly particular surgery questions even though you're not an MD like I thought you were*); Nana Shalikiani (*yes, we can go into Dr. Kuzanov's office and poke around on his computer while he's out of town*); Judy Berna (*yes, I'll be your friend and come with you to the amputee conference and introduce you all around*); Alexander Sah (*yes, you may spend the morning in my OR pestering my anesthesiologist and the unsuspecting sales rep from DePuy*); Jordan Newmark (*yes, I'll let you sit in on my fellowship seminar and teach you how to intubate someone even though you're not a fellow*); Mark Randolph (*yes, I'll welcome you into my home and set you up inside my wife's iron lung*); Hunter Cherwek and Kristin Taylor (*yes, you may come along to Mongolia for a week when we've never met you and have no idea where you're going with this*); Ed Pfueller (*yes, I'll speak openly with you and set you up in an ostomy run*); Adam Feinberg (*yes, I'll get everyone in my lab, plus me, to take time away from their research to help a know-nothing understand the complexities of printing with human cells*); Richard Chaffoo and Galina Thaler (*yes, I'll remove some follicles from your head and transplant one to your leg for no good reason*); Geoff Hamilton (*yes, you may spend two days at my biotech company wandering around laboratories full of trade secrets and delicate equipment*); Kevin D'Amour, Lisa McDonnell, and Antonella Pinto (*yes, I'll explain induced pluripotent stem cell science to you and I won't sigh audibly when you ask alarmingly basic questions*); José Luis Daza-Flores (*yes, I will chat candidly with you for hours while I liposculpt a torso*); TJ Roser and Susan Stuart (*yes, you may come watch a tissue recovery during a period when organ procurement organizations have been under scrutiny and there was every good reason to stay under the radar*). Thank you all.

My thanks also to the people who said yes to conversations—some in person, some by email, others on calls and Zooms that I promised would take ten minutes but rarely did: Wyeth Alexander, Branko Bojovic, Norma Braun, Peter Cataldo, David Chang, C. Lee Cohen, Shaoping Deng, Daniel Drake, Mona Eskandari, Don Gamble., W. Hock Hochheim, Kenichiro Imagawa, Ian Joesbury, Malik Kahook, Liangxue Lai, Lindsay Moore, Muhammad Mohiuddin, Janet Moradian-Oldak, Mike Olmes, Dengke Pan, Bill and Steven Raffensperger, Alvaro Rojas-Pena, Andrew Rosenbaum, Dave Rudzin, Julie Schallhorn, Tannin Schmidt, Jaimie Shores, Amelia Staats, Dawn Marie Sterling, the late Paul Stoodley, Bartek Szostakowski, Diana Tenney and Jerry Laperriere, and Takanori Takebe.

Acknowledgments

Early on, in the fullness of my ignorance, I picked some brains I've picked before: Leah Bellas, Justin Chickles, Anna Dhody, Mike Jones, Rick Redett, and Vail Reese. Thank you for opening the email even though you knew what was coming.

Chris Donhost, Ronni Dunn, Robert Emigh, and Sip Siperstein, thank you for the helpful connections and for vouching for me. Ann Eggold, Paul Mahon, and Elizabeth Roberts—you tried, and I deeply appreciate it. Eric Peterson, thanks for the books left outside the UCSF library when my access evaporated.

Books are a team sport. Without the contributions of some very talented and delightful people at W. W. Norton, I'd still be playing Little League. Jill Bialosky, my friend and editor since 2001, thank you for your wisdom, patience, and exceptional instincts. Erin Lovett, Meredith McGinnis, Brendan Curry, Ingsu Liu, Louise Brockett, Becky Homiski, Don Rifkin, you excel where I fall flat. Enduring gratitude for your support and inestimable gifts. Pat Holl, Laura Mucha, and Steph Romeo, I am indebted to you for your guidance, good humor, and help with photo permissions. Jay Mandel and Liv Guion at WME, you always have my back, and that is no small thing for an author. Janet Byrne, your title may be copyeditor, but savior is more apt. Lila Goodwillie, your patience and social media savvy have been a blessing and a boon. Buckets of gratitude delivered to your doors!

Alec, Kat, Malia, and the rest of my Notto-mates, thank you for keeping me sane. Steph Gold, dear friend and longtime travel pal, there are no words. Thank you for coming along. Ed, thank you for standing by me through the highs and lows and mid-altitude freak-outs of eight books. I love you with all my cardiomyocytes.

SOURCES

As is clear from the text, much of the information in this book comes from conversations with the people I spent time with or contacted by phone or email. What's below are citations that support other material. They're listed in order of the material's appearance in the chapter.

First Thoughts

Images of and information about masticators can be found in the catalogue of the Wellcome Collection and the March 21, 2012, blog post of the Museum of Health Care, in Kingston, Ontario. The relevant pages from *The Strange Story of False Teeth* (Routledge & Kegan Paul, 1983 reprint) are as follows: 4, 29, 46, 49, 66, 74 (early dentures), 100, 102 (George Washington's dental woes). The Poligrip survey is entitled "New U.S. Oral Care Survey Reveals Over Half (56%) of Denture Wearers Avoid Eating Foods; Also Feel Limitations in Work, Social and Romantic Lives"; it was posted on June 25, 2015, in the news section of GSK Consumer Healthcare.

The 1,000-plus reader responses to the r/AskHistorians subreddit about youthful denture wearers appear to have been deleted by a moderator, owing to the respondents' failure to "provide an academic understanding of the topic." Academic documentation can be found in Sara C. Gordon et al., "Prenuptial Extractions in Acadian Women: First Report of a Cultural Tradition," *Journal of Women's Health* 20(12): 1813–1818 (December 14, 2011); and Inkoo Kang and Julie M. Fagan, "Changing Dental Practices Amongst the Amish: Establishment of Dental Clinics to Treat Rather Than Yank Out Teeth," Amish Dental Intervention Project (2013).

Footnote page 5: Sergio Canavero, "Whole Brain Transplantation in Man: Technically Feasible," *Surgical Neurology International* 13: 594 (December 23, 2022). (Canavero is an associate editor in chief of the publication.)

1. To Build a Nose

I drew on two scholarly articles about Tycho Brahe's nose: Dennis C. Lee, "Tycho Brahe and His Sixteenth Century Nasal Prosthesis," *Plastic & Reconstructive Surgery* 50(4): 332–337 (October 1972), which includes the portrait with the exuberant mustaches, and Lene Østermark-Johansen's deeper dive, "The New Star, the New Nose: Tycho Brache's Nasal Prosthesis," *Nordic Journal of Renaissance Studies* 12: 93–105 (April 21, 2017). The biography mentioned on page 9 is by Grove Wilson, *Great Men of Science* (Garden City, NY: Garden City Publishing Company, 1932).

Frank Tetamore describes his spectacles-mounted nasal prostheses in "Deformities of the Face: Report of Plastic Operations and Also Mechanical Appliances for Covering Disfigurement," in Frank L. R. Tetamore, *Deformities of the Face and Orthopedics* (Brooklyn, NY: Press of the Adams Printing Co., 1894). Dr. Upham details his spring-loaded noses in "Artificial Noses and Ears," *Boston Medical and Surgical Journal* 145(19): 522–523 (November 7, 1901).

For the section on Indian and Italian nose reconstruction techniques: Gina C. Ang, "History of Skin Transplantation," *Clinics in Dermatology* 23(4): 320–324 (July–August 2005); David A. Shaye, "The History of Nasal Reconstruction," *Current Opinion in Otolaryngology and Head and Neck Surgery* 29(4): 259–264 (August 1, 2021); Charles M. Tipton, "Susruta of India, an Unrecognized Contributor to the History of Exercise Physiology," *Journal of Applied Physiology* 104(6): 1553–1556 (June 2008); Thomas Pennant, *The View of Hindoostan*, Volume II (London: Henry Hughs, 1798); Sophie Ménard, "An Unknown Renaissance Portrait of Tagliacozzi (1545–1599), the Founder of Plastic Surgery," *Plastic and Reconstructive Surgery—Global Open* 7(1): e2006 (January 4, 2019); Aina Greig et al., "Heinrich von Pfalzpaint, Pioneer of Arm Flap Nasal Reconstruction in 1460, More Than a Century Before Tagliacozzi," *Journal of Cranofacial Surgery* 26(4): 1165–1168 (June 2015).

Moving on to xenografting: Charles Sédillot, *Contributions à la Chirurgie*, Volume 2, page 589 (Paris: J. B. Baillière et Fils, 1868); Redard's chicken skin work is summarized in "Chicken-Skin for Grafting," Special Correspondence, *British Medical Journal* 2 (1453): 1018 (November 3, 1888); E. W. Lee's account of the pedicled pig graft is excerpted in Curtis P. Artz et al., "An Appraisal of Allografts and Xenografts as Biological Dressings for Wounds and Burns," *Annals of Surgery* 175(6): 934–938 (June 1972); Samuel W. Lambert, "The Vagaries of a Vivisectionist Turned Clinical Surgeon and the Story of the Lady Who Lay with a Pig for Five Nights and Five Days on Professional Advice," *Proceedings of the Charaka Club* IX: 38–43 (1938); "Remarkable Case of Skin Grafting," *Brooklyn Daily Eagle*, Sunday, September 18, 1898, page 28; Alexander Miles, "Notes on Skin Grafting from the Lower Animals," *Edinburgh Hospital Reports* 3: 647–661 (1895); M. E. Van Meter, "Note on the Use of Skin from Puppies in Skin-Grafting," *Annals of Surgery* 12: 136 (1890); G. F. Cadogan-Masterman, "Dermepenthesis," *British Medical Journal* 1(1413): 187 (January 28, 1888); Leonard Freeman, *Skin Grafting for Surgeons and General Practitioners* (St. Louis: C. V. Mosby Company, 1912), pages 85–88;

H. W. M. Kendall, "Grafting with Frog Skin," *British Medical Journal* 2(2915): 646 (November 11, 1916); "Frog Skin Used in Effort to Keep Texas Woman from Croaking," *Kansas City Journal,* March 21, 1907; "Skin of Thirty Frogs Is Grafted on Woman," Bedford, Indiana, *Times-Mail,* November 4, 1918, page 3; James E. Crowe Jr., "Treating Flu with Skin of Frog" *Immunity* 46(4): 517–518 (April 18, 2017); the account of Thomas Reardon is from "Frog Skin Leg," *Oakland Enquirer,* June 26, 1911, page 5; William Allen, "Skin Grafts from the Frog," letter to the editor in *The Lancet* 124(3194): 875–876 (November 15, 1884).

The details about the Cocoanut Grove burn patients are from Bradford Cannon, "Procedures in Rehabilitation of the Severely Burned," *Annals of Surgery* 117(6): 903–910 (1943).

Footnote page 11: Georgiana Macovei et al., "Changes in Dento-Facial Morphology Induced by Wind Instruments, in Professional Musicians and Physical Exercises That Can Prevent or Improve them—A Systematic Review," *Life (Basel)* 13(7): 1528 (July 8, 2023).

Footnote page 15: Examples of woodcuts can be found on page 421 of Leo M. Zimmerman and Katharine M. Howell, "History of Blood Transfusion," *Annals of Medical History* 4(5): 415–433 (September 1932); and page 24 of Robert A. Kilduffe and Michael DeBakey, *The Blood Bank and the Technique and Therapeutics of Transfusions* (St. Louis: C. V. Mosby Company, 1942). The quote is from "Transfusion of Lamb's Blood in the Human Subject," *Atlanta Medical and Surgical Journal* 12(10): 621–622 (January 1875), citing the *Medical Examiner.* "Direct Transfusion from Animal to Man," pages 234–235 of the *Atlanta Medical and Surgical Journal* 12(4), also citing the *Medical Examiner,* describe a lamb-to-man transfusion performed by Drs. Gadle, Prœgler, Hotz, and Wild (July 1874).

2. Gimme Some Skin

For background information on skin grafting, try Deepak Ozhathil et al., "A Narrative Review of the History of Skin Grafting in Burn Care," *Medicina* 57(4): 380 (April 15, 2021). You can read more about Diana Tenney and Jerry Laperriere's experiences in their memoir *God Never Moved: A Couple's Journey Through Fire to Life* (independently published, Amazon.com, 2023).

Footnote page 26: Frank Ashley, "Foreskins as Skin Grafts," *Annals of Surgery* 106(2): 252–256 (August 1937).

3. Mixed Meats

The details about China's high-rise piggeries are from two articles on the website Pig Progress (www.pigprogress.net): "Under Construction: A 26-Storey Pig House" and "Yangxiang Aims High with Sows on Many Floors." Lea Girani et al., "Xenotransplantation in Asia," *Xenotransplantation* 26(1): e12493 (2019). The information on China's use of death row prisoners is from "Organ Transplantation in China: Concerns Remain," *The Lancet* 385 (9971): 855–856 (March 7, 2015), which includes a reference to the *Beijing Times* interview.

Regarding miniature pigs in medical research: Dengke Pan et al., "Progress in Multiple Genetically Modified Minipigs for Xenotransplantation in China," *Xenotransplantation* 26(1): e12492 (2019). The relevant pages from "Swine in Biomedical Research," *Proceedings of a Symposium at the Pacific Northwest Laboratory,* July 19–22, 1965, Richland, Washington, ed. Leo K. Bustad and Roger O. McClellan, technical ed. M. Paul Burns, are 765 (history and overview), 307 (the Pavlov quote), 301 (pigs as a caricature of the obese and sedentary), 550 (pigs in braces).

The paper detailing the findings of the University of Alabama focus groups is Daniel J. Hurst et al., "Factors Influencing Attitudes Toward Xenotransplantation Clinical Trials," *Xenotransplantation* 28(4): e12684 (July 2021). The discussion of the ethics of chimeras can be found in Julian J. Koplin and Julian Savulescu, "Time to Rethink the Law on Part-Human Chimeras," *Journal of Law and the Biosciences* 6(1): 37–50 (May 15, 2019). Details on the kidney chimera are at Jiaowei Wang et al., "Generation of a Humanized Mesonephros in Pigs from Induced Pluripotent Stem Cells via Embryo Complementation," *Cell Stem Cell* 30: 1235–1245 (September 7, 2023). Yi Wang's work on encapsulation of porcine islet cells is described in Qi Zhang et al., "Islet Encapsulation: New Developments for the Treatment of Type 1 Diabetes," *Frontiers in Immunology* 13: 869984 (April 14, 2022).

4. Heart in a Box

For a detailed account of the Extracorporeal Life Support Laboratory's work extending the shelf life of hearts outside bodies, see Jennifer S. McLeod et al., "*Ex Vivo* Heart Perfusion for 72 Hours Using Plasma Cross Circulation," *ASAIO Journal* 66(7): 753–759 (July 2020). Bartlett's vitalin paper is from *Journal of the American College of Surgeons* 199(2): 286–292 (August 2004). For the information on ambulatory ECMO, see Bryan Boling et al., "Safety of Nurse-Led Ambulation for Patients on Venovenous Extracorporeal Membrane Oxygenation," *Progress in Transplantation* 26(2): 113–116 (June 2016).

Goat's milk transfusion references: John H. Brinton, "The Transfusion of Blood and the Intravenous Injection of Milk," *Medical Record* 14: 344–347 (1878); John Bryson, "Intravenous Injection of Milk in Chronic Bright's Disease—With Remarks," *St. Louis Courier of Medicine* 1: 523–528 (1879); T. Gaillard Thomas, "The Intra-Venous Injection of Milk as a Substitute for the Transfusion of Blood: Illustrated by Seven Operations," *New York Medical Journal* 27(5): 450–465 (May 1878); Edward M. Hodder, "Transfusion of Milk in Cholera," *Practitioner* 10: 14–16 (1873).

The investigator pursuing trials of enteral ventilation is Takanori Takebe, a professor with Tokyo Medical and Dental University and the Cincinnati Children's Hospital Medical Center. Takebe's company is EVA Therapeutics. The study "Enteral Liquid Ventilation Oxygenates a Hypoxic Pig Model," by T. Fujii et al., is in *iScience* 26(3): 106142 (February 13, 2023).

For more on Demetris Yannopoulos and mobile ECMO in emergency medicine, see Helen Ouyang, "The Race to Reinvent CPR," *New York Times Magazine*, March 27, 2024.

Footnote page 57: Gavitt A. Woodard and Georg M. Wieselthaler, "Cast of the Right Bronchial Tree," *New England Journal of Medicine* 379(22): 2151 (November 28, 2018).

Footnote page 58: Samir K. Ballas, "The First Cardeza Donor Center: Attracting Donors Who Do Not Wish to See Blood," *Transfusion* 53(1): 13 (January 2013).

Footnote page 59: Zimmerman and Howell, "History of Blood Transfusion."

Footnote page 66: Leland C. Clark Jr. and Frank Gollan, "Survival of Mammals Breathing Organic Liquids Equilibrated with Oxygen at Atmospheric Pressure," *Science* 152(3730): 1755–1756 (June 24, 1966); Da Jung Kim et al., "The 'Polka-Dot' Sign," *Abdominal Radiology* 42(8): 2194–2196 (August 2017).

5. The Vagina Dialogue

The *BioMed Research International* paper referenced on page 72 is Christopher J. Salgado et al., "Primary Sigmoid Vaginoplasty in Transwomen: Technique and Outcomes" (published online May 10, 2018). Dylan Isaacson et al., "How Big Is Too Big? The Girth of Bestselling Insertive Sex Toys to Guide Maximal Neophallus Dimensions," can be found in the *Journal of Sexual Medicine* 14(11): 1455–1461 (November 2017). For an illustrated overview of traditional vaginoplasty technique, try Poone Shoureshi and Daniel Dugi III, "Penile Inversion Vaginoplasty Technique," *Urologic Clinics of North America* 46(5): 511–525 (2019). The statistic on breast reconstruction complications is from Roni Caryn Rabin, "One in Three Women Undergoing Breast Reconstruction Have Complications," in the June 20, 2018, *New York Times*. The information about regulatory matters in the development of clinical procedures comes from "Technological Innovation: Comparing Development of Drugs, Devices, and Procedures in Medicine," by Annetine C. Gelijns and the U.S. Institute of Medicine Committee on Technological Innovation in Medicine (Washington, DC: National Academies Press, 1989).

Footnote page 76: Hiroshi Ikegaya et al., "Nose Size Indicates Maximum Penile Length," *Basic and Clinical Andrology* 31(1): 3 (February 4, 2021).

6. Giving the Finger

The translated title of Iva Kuzanov's paper is "Finger Transplant in the Creation and Reconstruction of the Penis," *Khirurgiia* 10: 43–46 (2002). The A. P. Frumkin orgasm quote is from the last page of "Reconstruction of the Male Genitalia," from page 14 of Volume 2 of *American Review of Soviet Medicine* (1944), ed. Henry E. Sigerist.

Discussion of the challenges of composite tissue transplants comes mainly from a phone conversation with Branko Bojovic. For more, see P. C. Cavadas et al., "Bilateral Transfemoral Lower Extremity Transplantation: Result at 1 Year,"

American Journal of Transplantation 13(5): 1343–1349 (February 22, 2013); Maria Siemionow, "The Miracle of Face Transplantation After 10 Years," *British Medical Bulletin* 120(1): 5–14 (October 28, 2016).

Footnote page 86: Antonio Perciaccante et al., "Gangrene, Amputation, and Allogeneic Transplantation in the Fifth Century AD: A Pictorial Representation," *Journal of Vascular Surgery* 64(3): 824–825 (2016).

7. The Cut-Off Point

Judy Johnson Berna's memoir, *Just One Foot* (CreateSpace, 2012), conveys in detail the struggles involved in voluntarily parting with a limb. G. Howson (ed.) and John Galsworthy (foreword), *Handbook for the Limbless* (London: The Disabled Society, 1921); K. J. Kinch and J. D. Clasper, "A Brief History of War Amputation," *Journal of the Royal Army Medical Corps* 157(4): 374–380 (December 2011).

Some studies comparing outcomes with amputation versus with limb salvage: Levent Tekin et al., "Comparison of Quality of Life and Functionality in Patients with Traumatic Unilateral Below Knee Amputation and Salvage Surgery," *Prosthetics and Orthotics International* 33(1): 17–24 (March 2009); Margaret J. Fairhurst, "The Function of Below-Knee Amputee Versus the Patient with Salvaged Grade III Tibial Fracture," *Clinical Orthopaedics and Related Research* 301: 227–232 (April 1994); Isaac Okereke and Elsenosy Abdelfatah, "Limb Salvage Versus Amputation for the Mangled Extremity: Factors Affecting Decision-Making and Outcomes," *Cureus* 14(8): e28153 (August 18, 2022); Kaj Johansen et al., "Objective Criteria Accurately Predict Amputation Following Lower Extremity Trauma," *Journal of Trauma* 30(5): 568–573 (May 1990).

Britt H. Young's essay "I Have One of the Most Advanced Prosthetic Arms in the World—and I Hate It" appeared on Inverse.com, March 4, 2021. See also her essay "My Body Is Used to Design Military Tech," Wired.com (October 26, 2021). The statement about upper arm prosthetic acceptance over time is from Stefan Salminger et al., "Current Rates of Prosthetic Usage in Upper-Limb Amputees—Have Innovations Had an Impact on Device Acceptance?" *Disability and Rehabilitation* 44(14): 3708–3713 (July 2022).

For the osseointegration section: Jason Shih Hoellwarth et al., "The Clinical History and Basic Science Origins of Transcutaneous Osseointegration for Amputees," *Advances in Orthopedics* (March 18, 2022); Kerstin Hagburg et al., "Osseointegrated Prostheses for the Rehabilitation of Patients with Transfemoral Amputations: A Prospective Ten-Year Cohort Study of Patient-Reported Outcomes and Complications," *Journal of Orthopedic Translation* 38: 56–64 (2023); Charlotte Grieve, Tom Steinfort, and Natalie Clancy, "Oozing and Maggots: The Stories One of Australia's Most Celebrated Surgeons Doesn't Want You to Hear," *Sydney Morning Herald*, September 18, 2022; Grant G. Black et al., "Osseointegration for Lower Limb Amputation: Understanding the Risk Factors and Time Courses of Soft Tissue Complications," *Annals of Plastic Surgery* 90(6), Supplement 5: S452–S456; Christopher Rennie et al., "Complications Following

Osseointegrated Transfemoral and Transtibial Implants: A Systematic Review," *Cureus* 16(3): e57045 (March 2024).

Footnote page 92: Anand N. Bosmia, Christoph J. Griessenauer, and R. Shane Tubbs, "Yubitsume: Ritualistic Self-Amputation of Proximal Digits Among the Yakuza," *Journal of Injury and Violence Research* 6(2): 54–56 (July 2014).

Footnote page 98: Caroline Leech and Keith Porter, "Man or Machine? An Experimental Study of Prehospital Emergency Amputation," *Emergency Medicine Journal* 33(9): 641–644 (June 2016).

Footnote page 98: Mark Skippen et al., "The Chain Saw: A Scottish Invention," *Scottish Medical Journal* 49(2): 72–75 (May 2004).

8. Joint Ventures

Paul Anderson's woodworking class is described in Biloine W. Young, "The Carpenter Surgeon, Literally," *Orthopedics This Week,* March 19, 2011. Information on sales reps in the OR is from pages 75–78 of Julie Anderson, Francis Neary, and John V. Pickstone, *Surgeons, Manufacturers, and Patients: A Transatlantic History of Total Hip Replacement* (New York: Palgrave Macmillan, 2007). For a technical discussion of the effects of wear debris, see Diana Bitar and Javad Parvizi, "Biological Response to Prosthetic Debris," *World Journal of Orthopedics* 6(2): 172–189 (March 18, 2015). Debra Cohen's investigation of the ASR fiasco is at "Out of Joint: The Story of the ASR," *BMJ* 342: d2905 (May 13, 2011). Christopher A. Jarrett et al., "The Squeaking Hip: A Phenomenon of Ceramic-on-Ceramic Total Hip Arthroplasty," *Journal of Bone and Joint Surgery* 91(6): 1344–1349 (June 2009). John Charnley's biography is by William Waugh, *John Charnley: The Man and the Hip* (London: Springer-Verlag, 1990). Charnley and Nas S. Eftekhar's paper "Penetration of Gown Material by Organisms from the Surgeon's Body" is from *The Lancet* 293(7587): 172–173 (January 25, 1969).

The statistic on infection and revision surgery is from the American Joint Replacement Registry's 2023 annual report. A conversation with Bartek Szostakowski yielded much of the story of San Baw's ivory prostheses. I also made use of his papers, including Bartek Szostakowski and John A. Skinner, "Dr. San Baw: The Man Behind Ivory Hemiarthroplasty," *BJJ News* 5 (December 2014); Szostakowski et al., "ArtiFacts: Ivory Hemiarthroplasty: The Forgotten Concept Lives on," *Clinical Orthopaedics and Related Research* 475(12): 2850–2854 (December 2017); and Bartek Szostakowski and Marlene DeMaio, "Ideal Xenograft or a Perfect Bone Substitute?—A Retrospective Review and Analysis of the Historical Concept of Ivory Implants in Orthopaedics," *International Orthopedics* 44(5): 1003–1009 (May 2020). See also Daniel Stiles, "Ivory Carving in Myanmar," asianart.com, November 19, 2002.

The information on biofilm formation comes from a conversation with Paul Stoodley and Amelia Staats, as well as their paper, with Daniel Li and Anne C. Sullivan, "Biofilm Formation in Periprosthetic Joint Infections," *Annals of Joint* 6: 43 (October 15, 2021). The Stoodley Lab study of biofilms on joint implants is Kelly Moore et al., "Mapping Bacterial Biofilm on Features of

Orthopedic Implants In Vitro," *Microorganisms* 10: 586 (March 8, 2022). Also see Stefanie M. Shiels et al., "Revisiting the 'Race for the Surface' in a Pre-Clinical Model of Implant Infection," *European Cells & Materials* 39: 77–95 (January 29, 2020); and Shan S. Lansing, "High Number of Door Openings Increases the Bacterial Load of the Operating Room," *Surgical Infections* 22(7): 684–689 (September 2021).

The statistic on the prevalence of infection after hip replacement surgery is from Dr. Sah, verified in Rajesh Malhotra et al., "Arthroplasty-Associated Infections," Medscape.com, August 10, 2023. For anyone considering a hip or knee replacement, the annual report of the American Academy of Orthopaedic Surgeons' American Joint Replacement Registry is an invaluable resource, providing clear, detailed data on trends in materials and surgical approaches, as well as on complications.

Footnote page 117: Austin T. Moore, "Metal Hip Joint: A New Self-Locking Vitallium Prosthesis," *Southern Medical Journal* 45(11): 1015–1019 (November 1952).

Footnote page 119: Nikitas N. Nomikos and Chris K. Yiannakoupolos, "The First Shoulder Replacement in Ancient Greek Mythology: The Story of Pelops, King of Elis," *Orthopaedics & Traumatology: Surgery & Research* 105(5): 801–803 (September 2019).

9. Intubation for Dummies

The statistic on death during intubation is from an article by Zahid Hussain Khan et al., "Mortality Related to Intubation in Adult General ICUs: A Systematic Review and Meta-Analysis," *Archives of Neuroscience* 7(3): e89993 (July 14, 2020).

Footnote page 131: Harry M. Sherman, "The Green Operating Room at St. Luke's Hospital," *California State Journal of Medicine* 12(5): 181–183 (May 1914).

10. Heavy Breathing

Dr. Lewins's "Apparatus for Promoting Respiration in Cases of Suspended Animation" is a letter to the editor of the *Edinburgh Medical and Surgical Journal* 54: 255–256 (1840). For the history of negative-pressure ventilators, I relied on Norma M. T. Braun's "Negative Pressure Noninvasive Ventilation (NPNIV): History, Rationale, and Application," which is chapter 2 of *Nocturnal Non-Invasive Ventilation: Theory, Evidence, and Clinical Practice*, ed. Robert C. Basner and Sairam Parthasarathy (New York: Springer Science and Business Media, 2015); as well as the booklet "The Evolution of Iron Lungs" (Cambridge, MA: J. H. Emerson Company, 1958).

The polio memoirs I reference are Regina Woods, *Tales from Inside the Iron Lung (and How I Got Out of It)* (Philadelphia: University of Pennsylvania Press, 1994); Kathryn Black, *In the Shadow of Polio: A Personal and Social History* (New York: Addison-Wesley, 1996); Arnold R. Beisser, *Flying Without Wings: Personal Reflections on Being Disabled* (Boston: G. K. Hall & Co., 1992). Also helpful was the pamphlet "The Nursing Care of the Patient in the Respirator," by Carmelita Calderwood (National Foundation for Infantile Paralysis, 1944).

Back issues of the *Toomeyville Jr. Gazette* are available on the website Polio-Place.org (http://polioplace.org/GINI). The website of the Disability History Museum (www.disabilitymuseum.org) is another good resource for historical information on iron lung use during the polio era.

The statistic on CPAP use is from Norman Wolkove et al., "Long-Term Compliance with Continuous Positive Airway Pressure in Patients with Obstructive Sleep Apnea," *Canadian Respiratory Journal* 15(7): 365–369 (October 2008). Regarding ICU dental woes, I used François Fourier et al., "Colonization of Dental Plaque: A Source of Nosocomial Infections in Intensive Care Unit Patients," *Critical Care Medicine* 26(2): 301–308 (February 1998); and Mi-Kyoung Jun et al., "Hospital Dentistry for Intensive Care Patients: A Comprehensive Review," *Journal of Clinical Medicine* 10(16): 3681 (August 19, 2021). The statistic on deaths from ventilator-associated pneumonia comes from the latter. See also Mark J. Butler et al., "Mechanical Ventilation for COVID-19: Outcomes Following Discharge from Inpatient Treatment," *PLoS ONE* (January 6, 2023). Mona Eskandari's project is described in Samaneh Sattari et al., "Introducing a Custom-Designed Volume-Pressure Machine for Novel Measurements of Whole Lung Organ Viscoelasticity and Direct Comparisons Between Positive- and Negative-Pressure Ventilation," *Frontiers in Bioengineering and Biotechnology* 8: 578762 (October 21, 2020). The statistic on hearing loss is from Lauren K. Dillard et al., "Prevalence of Aminoglycoside-Induced Hearing Loss in Drug-Resistant Tuberculosis Patients: A Systemic Review," *Journal of Infection* 83(1): 27–36 (2021).

The website of the charity Exovent (www.exovent.org) is a good overall source of information about the advantages of negative-pressure ventilation.

Footnote page 140: Linda Wang, "Why Are Trick Candle Flames So Impossible to Blow Out?," *Chemical & Engineering News* 88(32): 34 (August 9, 2010).

11. The Mongolian Eyeball

For the historical aspects of cataract surgery, I relied on Robert Henry Elliot, *The Indian Operation of Couching for Cataract* (New York: Paul B. Hoeber, 1918), and the late Charles D. Kelman's memoir, *Through My Eyes: The Story of a Surgeon Who Dared to Take on the Medical World* (New York: Crown Publishers, 1985). Kelman's description of the various rotating devices used in his early efforts to remove cataracts are in the September 2006 Up Front section of CRST Europe, under the title "The Genesis of Phacoemulsification." The statistic on blindness from cataract surgery is from Shaozhen Li et al., "A Survey of Blindness and Cataract Surgery in Doumen County, China," *Ophthalmology* 106(8): 1602–1608 (1999).

The report on accommodating lenses is Julie M. Schallhorn et al., "Multifocal and Accommodating Intraocular Lenses for the Treatment of Presbyopia: A Report by the American Academy of Ophthalmology," *Ophthalmology* 128(10): 1469–1482 (October 2021).

Footnote page 160: Howard C. Howland, Monica Howland, and Christopher J. Murphy, "Refractive State of the Rhinoceros," *Vision Research* 33(18): 2649–2651 (December 1993).

12. The Last Six Inches

The Heister and Pillore quotes appear in Peter A. Cataldo's "Intestinal Stomas: 200 Years of Digging," *Diseases of the Colon & Rectum* 42(2): 137–142 (February 1999). Daniel Pring's case study of Mrs. White is "History of a Case of the Successful Formation of an Artificial Anus in an Adult," *London Medical and Physical Journal* 45(263): 1–15 (January 1821). Richard Martland describes the Henry Baron case in "Case in Which the Operation for Artificial Anus was Successfully Performed," *Edinburgh Medical and Surgical Journal* 24(85): 271–276 (October 1, 1825), with a sequel January 1, 1830, issue 33(102): 80–82. The fatwa discussed on page 181 can be found, in Arabic and in English, on the website of the International Ostomy Association (http://ostomyinternational.org/fatwa/). The petition for an apology from the Cincinnati Police Department can be found on change.org, "Apologize for Their Inadvertent Discrimination Towards Ostomy Patients," started by Douglas Yakich, July 31, 2013.

For the section on artificial and lab-grown anal sphincters: D. F. Altomare et al., "Disappointing Long-Term Results of the Artificial Anal Sphincter for Faecal Incontinence," *British Journal of Surgery* 91(10): 1352–1353 (October 2004); the infection rate statistic is from Sharon G. Gregorcyk, "The Current Status of the Acticon® Neosphincter," *Clinics in Colon and Rectal Surgery* 18(1): 32–37 (February 2005); the rabbit paper is Jaime Bohl, Elie Zakhem, and Kalil N. Bitar, "Successful Treatment of Passive Fecal Incontinence in an Animal Model Using Engineered Biosphincters: A 3-Month Follow-Up Study," *Stem Cells Translational Medicine* 6(9): 1795–1802 (September 2017); the primate study is Prabhash Dadhich et al., "BioSphincters to Treat Fecal Incontinence in Nonhuman Primates," *Scientific Reports* 9(1): 18096 (December 2019). The statistic on prevalence of fecal incontinence is from Nallely Saldana-Ruiz and Andreas M. Kaiser, "Fecal Incontinence: Challenges and Solutions," *World Journal of Gastroenterology* 23(1): 11–24 (January 2017).

Footnote page 175: Alessandro Lugli et al., "The Gastric Disease of Napoleon Bonaparte: Brief Report for the Bicentenary of Napoleon's Death on St. Helena in 1821," *Virchows Archiv* 479: 1055–1060 (March 4, 2021); Arline Meyer, "Re-dressing Classical Statuary: The Eighteenth-Century 'Hand-in-Waistcoat' Portrait," *Art Bulletin* 77(1): 45–63 (March 1995).

13. Out of Ink

Here's the paper about the Wake Forest bladders: Anthony Atala et al., "Tissue-Engineered Autologous Bladders for Patients Needing Cystoplasty," *The Lancet* 367: 1243–1246 (published online April 16, 2006). For a more recent summary of the challenges of lab-grown bladders, see Stephanie L. Osborn and Eric

A. Kurzrock, "Bioengineered Bladder Tissue—Close but Yet So Far!" *Journal of Urology* 194(3): 619–620 (September 2015).

14. Shaft

The story of Norman Orentreich's discovery being booed at a meeting was related to me by Richard Chaffoo. Norman Orentreich, "Autografts in Alopecias and Other Selected Dermatological Conditions," *Annals of the New York Academy of Sciences* 83: 463–479 (November 1959). The Osakan physicians I reference are Shigeki Inui and Satoshi Itami, whose letter to the editor "Dr. Shoji Okuda (1886–1962): The Great Pioneer of Punch Graft Hair Transplantation" ran in *Journal of Dermatology* 36(10): 561–562 (October 2009). Kenichiro Imagawa and Shigeki Inui, "A Visit to the House of Dr. Shoji Okuda," *Hair Transplant Forum International* 19(6): 201–202 (November/December 2009). An English translation of the Okuda papers can be accessed on the website of the International Society of Hair Restoration Surgery.

Sanusi Umar describes his study in "Body Hair Transplant by Follicular Unit Extraction: My Experience with 122 Patients," *Aesthetic Surgery Journal* 36(10): 1101–1110 (November 2016). The paper referenced on page 206 is Young-Ran Lee et al., "Hair Restoration Surgery in Patients with Pubic Atrichosis or Hypotrichosis: Review of Technique and Clinical Considerations of 507 Cases," *Dermatological Surgery* 32(11): 1327–1335 (November 2006). The statistic on survival of scalp hair transplanted to the leg is from page 1334 of that paper.

Footnote page 206: Philip R. Cohen, "Scrotal Rejuvenation," *Cureus* 10(3): e2316 (March 13, 2018).

15. Splitting Hairs

The information in this chapter comes from two visits to Stemson Therapeutics, during which Kevin D'Amour, Lisa McDonnell, and Antonella Pinto patiently explained the basics of inducing adult cells to regress into pluripotent stem cells and then guiding them to become other types of cells. As well as other topics in regenerative medicine and hair regrowth. Denise Grady, "Patients Lose Sight After Stem Cells Are Injected into Their Eyes," *New York Times*, March 15, 2017.

Footnote page 219: Erin Orozco, Alysa Birnbrich, and Shari R. Liberman, "The Role of N-Acetylcysteine in the Treatment of Accidental Submersion of the Hands in Liquid Nitrogen," *Cureus* 13(9): e18129 (September 2021).

Footnote page 221: Aditya K. Gupta, Maanasa Venkataraman, and Emma M. Quinlan, "Artificial Hair Implantation for Hair Restoration," *Journal of Dermatological Treatment* 33(3): 1312–1318 (May 2022); see also "Biofibre Hair Transplants: Everything You Need to Know," from the website of the Wimpole Clinic (which does not offer the procedure).

16. The Ass Men

Nikolas V. Chugay, Paul N. Chugay, and Melvin A. Shiffman, *Body Sculpting with Silicone Implants* (Heidelberg: Springer, 2014)—also the textbook referenced on page 227; Ramón Cuenca-Guerra and Jorge Quezada, "What Makes Buttocks Beautiful? A Review and Classification of the Determinants of Gluteal Beauty and the Surgical Techniques to Achieve Them," *Aesthetic Plastic Surgery* 28(5): 340–347 (September/October 2004); Mario González-Ulloa, "Gluteoplasty: A Ten-Year Report," *Aesthetic Plastic Surgery* 15: 85–91 (December 1991); Ramón Cuenca-Guerra, José Luis Daza-Flores, and Ale Jalil Saade-Saade, "Calf Implants," *Aesthetic Plastic Surgery* 33: 505–513 (July 2009); Lars Johan Sandberg et al., "Beyond the 21-cm Notch-to-Nipple Myth: Golden Proportions in Breast Aesthetics," *Plastic and Reconstructive Surgery Global Open* 9(10): e3826 (October 25, 2021); William M. Cocke and Greer Ricketson, "Gluteal Augmentation," *Plastic and Reconstructive Surgery* 52(1): 93 (July 1973); Patrick P. Bletsis et al., "Determining Breast Volume Preference Among Patients, Plastic Surgeons, and Laypeople: Is There a Perfect Breast Size?" *Journal of Plastic, Reconstructive & Aesthetic Surgery* 75(9): 3078–3084 (September 2022). The implant weights come from manufacturers' product sheets and from Ken Dembny, "How Much Does a Breast Implant Weigh?," July 4, 2014, on the website MyBreastAugmentation.info.

Here are some of the relevant papers on the golden ratio in human anatomy: Sulcuk Ozturk, Kenan Yalta, and Ertan Yetkin, "Golden Ratio: A Subtle Regulator in Our Body and Cardiovascular System?," *International Journal of Cardiology* 223: 143–145 (November 15, 2016); Alan Hutchison and Richard L. Hutchison, "Fibonacci, Littler, and the Hand: A Brief Review," *Hand* 5(4): 364–368 (April 2010); N. R. Vadachkoriia, N. Sh. Gumberidze, and N. A. Mandzhavidze, "'Golden Proportion' and Its Application to Calculate Dentition," *Georgian Medical News* 142: 87–94 (January 2007). More at H. E. Huntley, *The Divine Proportion: A Study in Mathematical Beauty* (New York: Dover Publications, 1970).

The information on breast augmentation via chunks of fat is from John Watson, "Some Observations on Free Fat Grafts: With Reference to Their Use in Mammaplasty," *British Journal of Plastic Surgery* 12: 263–274 (1959–1960); and J. P. Lalardrie and Roger Mouly, "History of Mammaplasty," *Aesthetic Plastic Surgery* 2(1): 167–176 (December 1978). M. Sharon Webb's 1997 dissertation "Cleopatra's Needle: The History and Legacy of Silicone Injections" can be accessed via Harvard University's DASH Repository (https://dash.harvard.edu/handle/1/8889460).

Footnote page 227: González-Ulloa's charcoal nudes (with disembodied hand) can be found in "Torsoplasty," *Aesthetic Plastic Surgery* 3: 357–368 (December 1979) and "Gluteoplasty: A Ten-Year Report," *Aesthetic Plastic Surgery* 15: 85–91 (Winter 1991).

Footnote page 228: Wendy W. Wong et al., "Redefining the Ideal Buttocks: A Population Analysis," *Plastic and Reconstructive Surgery* 137(6): 1739–1747 (June 2016).

Footnote page 231: The quote about Cronin's inspiration is from Anand K.

Deva et al., "The 'Game of Implants': A Perspective on the Crisis-Prone History of Breast Implants," *Aesthetic Surgery Journal* 39(s1), Supplement 1: S55–S65 (January 2019). The story also appears in Manish C. Champaneria et al., "The Evolution of Breast Reconstruction: A Historical Perspective," *World Journal of Surgery* 36: 730–742 (February 14, 2012), which is where I found the additional details about the dogs with implants. The detail about the sizes is from Thomas D. Cronin and Raymond O. Brauer, "Augmentation Mammaplasty," *Surgical Clinics of North America* 51(2): 441–449 (April 1971).

17. Some of the Parts

The title of the U.S. Department of Health and Human Services audit report on OPOs is *Medicare Paid Independent Organ Procurement Organizations Over Half a Million Dollars for Professional and Public Education Overhead Costs That Did Not Meet Medicare Requirements* (A-09-21-03020, August 2023). Andrew J. Rosenbaum and Timothy Tian Roberts's paper "Bone Grafts, Bone Substitutes and Orthobiologics," is from the October/November/December 2012 *Organogenesis*, pages 114–124. My cadaver goose pimple source is from Andrew M. Baker, chief medical examiner for Hennepin County, Minnesota.

Footnote page 248: Livia Francine Soriano, "Omphaloliths: A Case Series and Review of 29 Cases in Literature," *Dermatology Online Journal* 25(9): 13030 (September 15, 2019).

Last Thoughts

The two studies mentioned in the tears section are as follows: Tannin A. Schmidt et al., "Transcription, Translation, and Function of Lubricin, a Boundary Lubricant, at the Ocular Surface," *JAMA Ophthalmology* 131(6): 766–776 (June 2013); Leying Wang et al., "Autologous Serum Eye Drops versus Artificial Tear Drops for Dry Eye Disease: A Systematic Review and Meta-Analysis of Randomized Controlled Trials," *Ophthalmic Research* 63(5): 443–451 (2020).